CRITIS Series in Critical Thinking

First published 2024

Front Cover
The Thinker of Hamangia
Hamangia culture, Romania, around 5,000 BC. Adapted from: https://en.wikipedia.org/wiki/Thinker_of_Hamangia#/media/File:Bucharest_-_The_Thinker_&_The_Sitting_Woman_of_Cernavoda_-_no_bg.png which is used under CC BY-SA 4.0 License: https://creativecommons.org/licenses/by-sa/4.0/

Library of Congress Cataloging-in-Publication Data
Tecuci, Gheorghe
Critical Thinking: A Gentile Introduction

Identifiers: ISBN 978-1-964262-01-7 (hardcover)
ISBN 978-1-964262-00-0 (softcover)
ISBN 978-1-964262-02-4 (ebook)

Subjects: Critical Thinking. | Evidence. | Evidence-based Reasoning.| Artificial Intelligence. |

Classification:

ISBN 978-1-964262-00-0

CRITIS PRESS, Fairfax, VA, USA
criticalthinkingsystems@gmail.com

To Victor Davis Hanson

FOREWORD

William Twining
University College London
Fellow of the British Academy and the Academy for Social Sciences
Foreign Member of the American Academy of Arts and Sciences

This is a remarkable book. The author is a leading computer scientist with many theoretical and practical interests. These include, among others, artificial intelligence, reasoning in legal contexts, expert systems, and theoretical and practical contributions to intelligence analysis. He is passionate about education or, as he prefers to call it, learning. He has been a close follower, then collaborator, and now the successor of David Schum, a statistician and psychologist whose Evidential Foundations of Probabilistic Reasoning (1994/2001) is the starting point for many academics of modern efforts to develop Evidence as a multidisciplinary field in Law and in 'evidence-based' approaches more generally.

There is a vast literature on overlapping 'subjects' such as 'Critical Thinking,' 'Clear Thinking,' 'Problem-Solving,' 'Informal logic,' and 'Practical Reasoning.' This vastness is not least because they have been considered foundational in education since the medieval Trivium of Logic, Grammar, and Rhetoric. Some subjects now seem to be becoming overshadowed by the extraordinary rise of Artificial Intelligence. Tecuci, himself a leader in the field, steers a firm path between current Enthusiasts and Sceptics.

'Critical Thinking' is attributed to John Dewey (1910). Because of its broad reach, it has unsurprisingly been stretched in several directions. The breadth of this label is illustrated by a short list of leading thinkers associated with it: for an older generation (myself included), this list included George Polya, Herbert Simon, Susan Stebbing, Max Black, and Stephen Toulmin; more recently, they have been joined by Sharon Bailin, Douglas Walton, and, perhaps less obviously, Susan Haack. Many leading educationalists and statisticians could be added to such a list, together with David Schum and now Gheorghe Tecuci.

Given the variety of potential readers, this book can anticipate a variety of reactions, some enthusiastic, others critical. In my view, it deserves serious attention. I think that it is remarkable in four specific ways:

1. It brings together familiar topics with others not usually associated with 'critical thinking' such as intelligence analysis, modified Wigmorean approaches, and reasoning in legal contexts.

2. By rooting the approach in inferential reasoning, it focuses on the evidence-based movement in the health sciences and beyond and on efforts to develop Evidence as a distinctive multi-disciplinary field within and beyond Law (https://evidencedialogues.wordpress.com/).

3. The decision to use illustrations drawn from a single field works well: intelligence analysis is itself fascinating; the presentation combines theory and practice from a leading expert who has trained and advised the American intelligence community; there is a direct link with a more detailed book by the same author (Tecuci, Schum, et al. 2016); the examples feel more realistic than concocted hypotheticals; and they accumulate, enabling quite sophisticated treatment of concrete issues. Most readers of stories of espionage and criminal investigation, whether fictional or reported, will have a relevant background; some may even ponder whether there is a close affinity between Gheorghe Tecuci and George Smiley.

4. The book is also remarkable for its succinctness, especially as it links together so many topics hitherto perceived as diverse. Of course, succinctness courts many dangers. In this case, some of these are avoided because of this book's close links to more substantial texts by Schum, Tecuci, and others. The main achievement, in my view, is to subsume familiar and unfamiliar neighbors under a neat and persuasive structure. This is clearly illustrated by the Table of Contents, which is probably the best place for anyone interested to begin. It is an unrivaled overview of a vast landscape that has never before been mapped in this way.

Contents

PREFACE

Book Purpose

With the loss of trust in traditional news sources and the Internet, Google, YouTube, Facebook, Instagram, LinkedIn, personal Podcasts, and other individual and social media enterprises being the primary news sources, it became necessary for each of us to distinguish better between factual and fabricated information, not only in transnational events, such as the war between Russia and Ukraine or the barbaric attack of Israel by Hamas, but in every aspect of our lives, including politics, elections, work, and personal.

This textbook has been written as a gentle introduction to the complex critical thinking process, which refers to the ability to analyze information objectively and make reasoned judgments.

Conventional courses in logic, probability, and statistics do not prepare us for drawing conclusions based on masses of evidence whose items suggest many, often complex and interrelated, lines of arguments on hypotheses of interest. The evidence usually concerns unique, singular, or one-of-a-kind events and is thus not replicable or repeatable. This means there are rarely any applicable or relevant statistical records available to draw upon in making inferences about the capabilities and intentions of potential or actual adversaries. We had no existing statistical records regarding the intentions of foreigners who showed up in our civilian flying schools wishing only to learn how to steer multiengine aircraft and not perform takeoffs or landings.

Lacking existing statistical records to draw upon, we must generate new information by asking questions. Therefore, formal training is needed in the basic analytic tasks of developing novel and productive hypotheses and in constructing defensible and persuasive arguments based on masses of evidence.

We have arrived at several conclusions based on years of experience in training students, high-ranking military officers, and intelligence analysts. The first is that training in the evidential reasoning tasks required in these and other important contexts cannot be learned effectively just by listening to others discuss their analyses or by giving students lectures and assigned readings on the topics. Regular practice is necessary involving analyses of evidence using hypothetical or actual situations. Evidential analysis is mastered best by performing analyses contrived to illustrate the wide variety of subtleties or complexities often encountered in actual evidential analyses.

The second conclusion is based on our inspection of the materials offered in several courses for training intelligence analysts. It appears that analysts are so often trained in the production of intelligence analyses rather than in the actual analysis process itself. Very little training is offered regarding the properties, uses, discovery, and marshaling of the evidence upon which all analyses rest. As Douglas MacEachin (1999), former DDI, CIA, remarked:

> *"Information and expertise are a necessary but not sufficient means of making intelligence analysis the special product that it needs to be. A comparable effort has to be devoted to the science of analysis."*

The third conclusion is based on the strong emphasis in the Intelligence Community on developing computer-based tools to assist analysts. Indeed, analysts need all the help they can get in the face of a tsunami of information and the requirement to answer questions of immediate interest that do not allow time for extensive research and deliberation on the available evidence. Therefore, analysis tools based on solid theoretical foundations should be integrated into intelligence analysis courses to teach the analysis process (including the properties, uses, discovery, and marshaling of the evidence) through a learning-by-doing approach.

This textbook is written in a style congenial to students' interests regardless of their prior background and training. It will teach you basic knowledge about the properties, uses, discovery, and marshaling of evidence to show what is involved in assessing the evidence's relevance, credibility, and inferential force credentials. It includes a wide array of examples and exercises and the optional use of Cogent, an intelligent analytic tool involving real and hypothetical cases chosen to help you recognize and evaluate many of the complex elements of the analyses you are learning to perform.

Book Contents

Here is a route or map we will follow in your learning venture.

Chapter 1 introduces critical thinking, imaginative reasoning, and a systematic approach to critical thinking, illustrating them in intelligence analysis.

Chapter 2 discusses the types of questions we can ask, their possible answers, and the hypotheses to test.

Chapter 3 defines evidence and its three basic properties or credentials: relevance, credibility, and inferential force. These credentials are not inherent properties of evidence—they must be established through defensible and persuasive arguments.

Chapter 4 presents the evaluation of our sources of information, a highly complex process that used to be performed by news services that have now lost all credibility.

Chapter 5 discusses the five primary sources of uncertainty in masses of evidence: incompleteness, inconclusiveness, ambiguity, dissonance, and imperfect believability. This prepares the discussion of the development of arguments through an approach termed analysis and synthesis, divide and conquer, or task decomposition, in which we break some complex hypotheses into simpler elements and then reassemble these elements to form a conclusion.

Chapter 6 presents the methods for evidence-based hypothesis testing.

Chapter 7 illustrates the analysis of simple problems.

Chapter 8 discusses practical applications of critical thinking in intelligence, surveillance and reconnaissance, cybersecurity, anticipatory intelligence, malicious insider threats, law, forensics, and medicine.

We hope that your learning venture will be a valuable experience in which you will discover many important elements of critical thinking, some of which you might not have heard anything about before. We also hope this venture will directly relate to tasks you face or will face daily in your careers. Finally, we hope it will be as enjoyable as it is informative. So, as you begin this learning venture, we wish you ***Bon Voyage***!

Acknowledgments

David Schum (1932 – 2018) has significantly influenced our work on critical thinking and evidence-based reasoning.

We are grateful to Steven Rieber and IARPA for their support through the CREATE program, where the primary material for this book was developed. We also acknowledge the contributions of our collaborators in the CREATE program, particularly Louis Kaiser, Joan McIntyre, Richard Lempert, and Dorin Marcu.

We are also grateful to Susan Durham, William Nolte, John Donelan, Cindy Ayers, Jeffrey DeMatteis, and Jennifer Cassetti for their significant support and encouragement. Phillip Hwang, Joan Vallancewhitacre, John Greer, and Michelle Quirk supported our intelligence analysis research.

About the Author

Gheorghe Tecuci (Ph.Ds., University of Paris-Sud, now Paris-Sacley, and Polytechnic University of Bucharest, 1988) is Professor of Computer Science and Director of the Learning Agents Center in the School of Computing of George Mason University, Member of the Romanian Academy, and former Chair of Artificial Intelligence in the Center for Strategic Leadership of the U.S. Army War College. He has followed a career-long interest in developing a computational theory and technology that allows non–computer scientists to create cognitive agents that incorporate their problem-solving expertise and can act as cognitive assistants to experts, as expert consultants to nonexperts, and as intelligent tutors to students. He has published around 250 papers contributing to artificial intelligence, knowledge engineering, cognitive assistants, machine learning, evidence-based reasoning, and intelligence analysis. He has received the U.S. Army Outstanding Civilian Service Medal (for "groundbreaking contributions to the application of artificial intelligence to center of gravity determination") and the Innovative Application Award from the American Association for Artificial Intelligence (now the Association for the Advancement of Artificial Intelligence).

1 INTRODUCTION

1.1 Critical Thinking

Socrates

Teach thinking not knowledge.

Critical thinking is a complex concept that was developed over the past 2500 years through the work of some of the greatest minds, including Aristotle, Galileo Galilei, John Locke, Isaac Newton, William Whewell, Charles Peirce, John Wigmore, and David Schum, who have tried to understand the world through a process of discovery and testing of hypotheses based on evidence (see Figure 1). Critical thinking is the ability to objectively analyze information and make reasoned judgments. It is more precisely defined as the intellectually disciplined process of actively and skillfully conceptualizing, applying, analyzing, synthesizing, and evaluating information gathered from or generated by observation, experience, reflection, reasoning, or communication as a guide to belief and action (Scriven and Paul, 1987).

Critical thinking is incorporated into scientific, mathematical, historical, anthropological, economic, moral, and philosophical

Figure 1. Some of the greatest critical thinkers of all times.

thinking. It is at the core of problem-solving and decision-making tasks in various disciplines, including military science and intelligence, computing, natural and social sciences, education, agriculture, and medicine.

There are three essential components of critical thinking:

- *Hypotheses* (statements we would like to prove or disprove);
- *Evidence* (information used to prove or disprove the hypotheses);
- *Arguments* (logical links between our evidence and the hypotheses we entertain).

Consider the following situation:

> *A small bomb exploded in the city's center near a major university. No one was hurt, but a local monument sustained some damage. Initial police investigation indicates that the bomb had been hidden in a black backpack. Person X, a male student at the university and a member of an anti-government, anti-business anarchist group, has been identified as a person of interest. A witness has claimed to have seen Person X near the bombing site one hour before it occurred.*

Figure 2 shows an argument from the evidence. E^* (Person X was in the incident one hour before it occurred) to the hypothesis

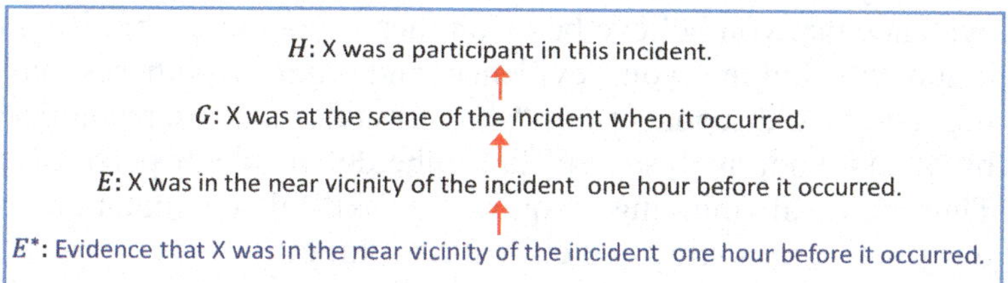

Figure 2. Argument from E^* to H.

H (Person X was a participant in this incident).

Our argument might run as follows:

- We have evidence that X was near the incident one hour before it occurred.
- Therefore, it is possible that X was near the incident one hour before it happened.

- Therefore, X might have been at the scene of the incident when it occurred.
- Therefore, if X was at the scene of the incident when it occurred, he might have been a participant in the incident.

1.2 Imaginative and Critical Reasoning

It would be a rare occurrence if you encountered an analytic task in which all possible hypotheses, all available evidence, and all arguments connecting the evidence and the hypotheses were supplied. You will have to generate or discover all these ingredients for yourself. This is where your imaginative reasoning becomes necessary.

Now, it happens that imaginative reasoning, though necessary, is not sufficient. Suppose you have generated some alternative hypotheses from the evidence you have discovered or selected from some larger collection of evidence that seems relevant to these hypotheses. You must also establish the relevance, credibility, and inferential force "credentials" of the evidence you have. This involves critical as well as imaginative reasoning on your part. You must be able to construct arguments from evidence to hypotheses that are both defensible and persuasive; this is where critical reasoning also becomes vitally necessary.

You may have generated entirely plausible hypotheses as well as evidence that you believe bears on these hypotheses, but if your arguments linking your evidence and your hypotheses have nonsequiturs, disconnects, or "short circuits" that are recognized by others, your analysis will fail to be defensible or persuasive. Thus, critical thinking requires a skillful combination of imaginative reasoning and critical reasoning.

1.3 Imaginative Reasoning

We often hear it claimed that some people have imaginative reasoning capabilities and others don't. If you don't have it, you are out of luck. The truth of the matter is that nature has endowed all of us with imaginative reasoning capabilities (Howe, 1999). The trouble is that we are not always given the opportunity or encouragement to be imaginative or creative in our thinking. Our work is based on the idea that you are naturally required to exercise your imagination to try to make sense of the masses of evidence you will encounter. Our role in this process is to assist you in various ways. What you think about the evidence you will discover is all-important. You may be able to assign possible meanings to evidence that others do not perceive.

Another fundamental matter concerns how productive the exercises of our imaginations are. We all encounter persons who seem to be imaginative in the new ideas they generate. However, many of these same persons do not always generate new ideas that are helpful in the analytic tasks at hand. So, what needs to be encouraged in critical thinking is productively imaginative thought. But there are other ingredients necessary to help you become more like Sherlock or Mycroft Holmes than Inspector Lestrade, the famous character of Conan Doyle.

You will be able to exercise your imaginative reasoning capabilities to the fullest only when you are driven by curiosity or wonder to find solutions to the analytic problems you encounter. If you don't care whether anyone finds a solution to these problems, you stand very little chance of generating a productively imaginative solution. Experience in many areas has shown that the most productively imaginative persons also have the greatest degree of commitment to finding solutions to problems that confront them.

The final ingredient we mention here concerns the diligence in how you approach each new analytic problem you face. There is an old saying that fortune favors the prepared mind. Unless you have done your homework in the particular substantive areas your analytical problems involve, you also stand little chance of generating productively imaginative new ideas. Your brain requires something to work with; as we all learn, this requires burning the midnight oil. However, being well acquainted only with the specifics of the substance of your analytic problems is often not quite enough.

Productively imaginative persons usually also have a breadth of knowledge and experience to draw upon. Productive new ideas often spring from the analogies we perceive; these analogies are often stated in metaphors. However, forming useful metaphors requires knowledge that goes beyond the boundaries of the believed substance of an analytic problem.

One of the most difficult problems we have faced is assisting you to construct defensible and persuasive arguments from a mass of evidence to hypotheses being considered. How well we can marshal our thoughts and evidence is vitally important in constructing defensible and persuasive arguments. The task of constructing arguments from a mass of different kinds of evidence is inherently difficult; perhaps it is the most difficult element of critical thinking.

Though methods for performing complex argument construction have been around for a long time, such as the Wigmorean methods (Wigmore 1937), few people have made particular use of them until quite recently. In this volume, we have combined concerns about these argument methods with concerns about thought and evidence marshaling.

A final point concerns the argument construction methods themselves. The methods we discuss may appear overly compulsive and may seem to require "too much thought." One response here is to remind persons reading our works that careful analyses always require careful thought, regardless of what methods are being used. Using the methods we describe, we construct "pictures" of a complex argument in the form of what today are called inference networks. You may have had some exposure to the use of various software systems that now exist for the probabilistic analysis of inference networks. The trouble is that no such system tells the user how to construct an inference network appropriate for the analysis of some existing mass of evidence. These systems all assume that the imaginative and critical reasoning steps necessary in inference network construction have already been performed by the user.

Having experience with the methods we discuss will offer analysts great assistance in seeing what is involved in the construction of defensible and persuasive arguments, regardless of whether you try to apply these methods in every analysis you undertake. Far too many people are looking for a book called

Critical Thinking Made Simple. We do not see any hope for any serious book or course with this title. Critical thinking is an inherently difficult task; the methods we describe form one way of coping with the complexity of such tasks.

1.4 The Discovery Method

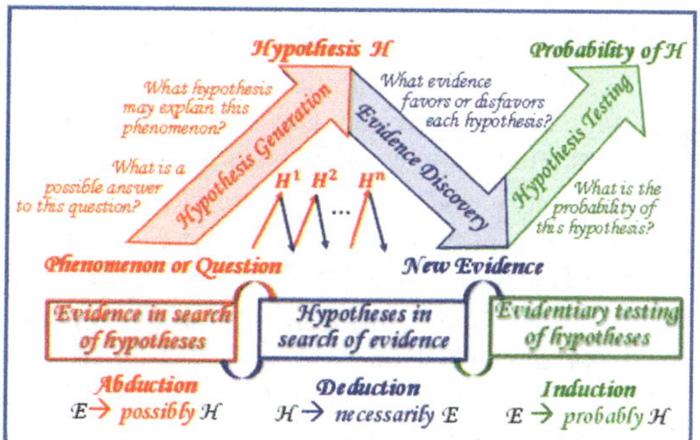

Figure 3. The discovery method.

Critical thinking may start with an interesting phenomenon to be explained or a question to be answered. The explanations of the phenomenon or the answers to the question are the hypotheses to be analyzed, and their identification involves abductive (imaginative) reasoning that shows something is possibly true.

To determine which of these hypotheses is true, one needs more evidence. Using deductive reasoning (showing something is necessarily true), one decomposes each hypothesis into simpler and simpler ones until the resulting hypotheses are simple enough to point to the evidence to look for.

Then, one uses inductive reasoning to show that something is probably true, to assess the probability of each hypothesis, and to determine the most likely hypothesis.

1.5 Critical Thinking in Intelligence Analysis

Intelligence analysis is the process of collecting information about an enemy to answer questions about its current operations or to predict its future behavior.

Consider the intelligence analysis problem from Table 1. Figure 3 is a representation of the reasoning to solve it.

First, we generate possible answers to the intelligence question:

Hakka has nuclear weapons.

Hakka does not have nuclear weapons.

Table 1. The Hakka problem.

Situation: Velovia, a technologically advanced democratic state, is home to Hakka, an apocalyptic sect suspected of plotting mass casualty terrorist attacks.

Question: Does Hakka have chemical weapons?

Available Information:

A source, who has reported accurately in the past, indicated that he knows a member of Hakka, Sulfurus, who has a bachelor's degree in chemistry.

According to official documents, Hakka has created two chemical companies. These companies have purchased technical equipment and substantial amounts of chemicals.

Regulatory and policing constraints are strong in Velovia. However, chemicals of the purity required for chemical weapons (e.g., sarin) can be procured at low visibility for plausibly legitimate business purposes.

A source from inside Hakka, who has reported accurately in the past, reported that Hakka can raise funds from its members, but the source was not specific on how much. The source also reported that Hakka does not have chemical weapons.

Hakka also runs a hospital that conducts some biological research for which it has purchased biological materials.

These are the hypotheses to be analyzed. They will be used to guide the discovery of relevant evidence. We decompose each hypothesis into simpler and simpler hypotheses until the resulting hypotheses are simple enough to show what evidence to look for. Consider, for example, the first hypothesis, which is that Hakka has nuclear weapons. One way to have nuclear weapons would be to develop them. This, in turn, would require expertise, production material, and funds (See Figure 4). "Hakka has the expertise to develop chemical weapons" is such a simple hypothesis for which we discover the following item of evidence:

> E1: A source who has reported accurately in the past indicated that Hakka has a member with a bachelor's degree in chemistry.

This information supports the truthfulness of the hypothesis "Hakka has the expertise to develop chemical weapons" and is, therefore, evidence for this hypothesis. In general, *evidence is any item of information that favors or disfavors the truthfulness of a hypothesis.*

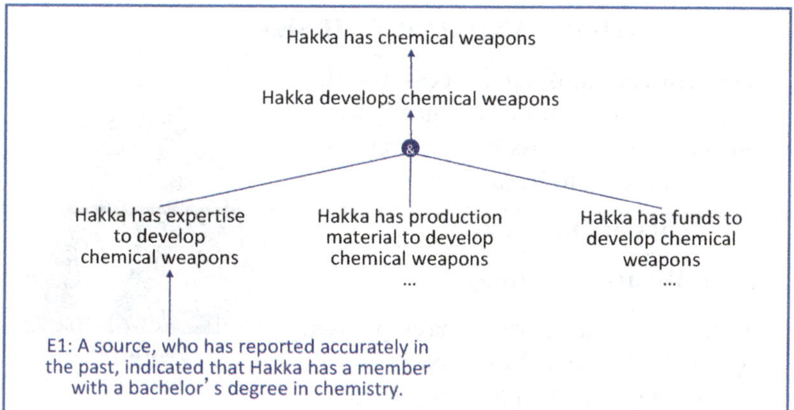

Figure 4. Evience discovery.

Finally, we employ inductive reasoning to determine the probability of the hypothesis, as shown in Figure 2.

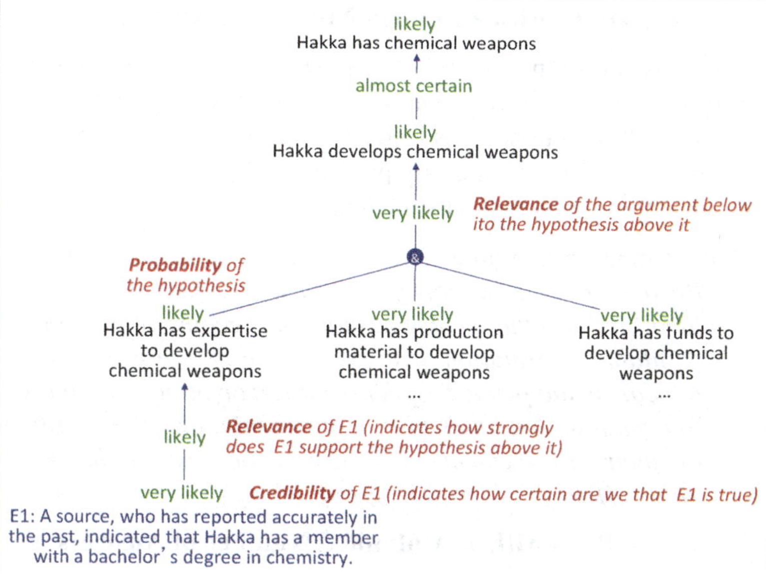

Figure 5. Hypothesis testing.

2 QUESTIONS AND HYPOTHESES

The critical thinking process usually starts with a question that needs to be answered. The possible answers are the hypotheses to be analyzed.

Confucius
The man who ask a question is a foul for a minute, the man who does not ask is a foul for life.

2.1 Types of Questions

2.1.1 Binary Questions

These are questions having two opposing answers, Yes or No:

- Does Hakka have chemical weapons?
- Is President Cardoza involved in the corruption scheme?
- Did John Steal the car?

2.1.2 Questions with a Limited Number of Answers

These are questions having a finite set of answers determined by a finite set of possibilities based on prior knowledge, such as:

- Which missile was tested?
- Which terrorist group performed the terror attack?
- Who passed the documents?

Here is an important note concerning this type of question:

Because of information gaps, all of the potential answers identified in the problem may be inconsistent with the available information. For example, if two terrorist groups appear as the prime suspects in a terror attack, but there is information that tends to rule out or minimize the likelihood of their involvement, the correct answer might be an unidentified third terrorist group.

2.1.3 Questions with an Unlimited Number of Answers

These are questions having a potentially unlimited set of answers that ultimately is constrained by the available information, such as:

- Why did the foreign minister resign?
- Why did President Marcos reject the treaty?
- Why is country Redland developing nuclear weapons?

2.1.4 Review Questions

1. Characterize each of the questions below concerning the number of answers it can have.
   - Will the president select General Martin to be the country's next defense minister?
   - Which of the country's four-star generals is the president likely to nominate as the country's next defense minister?
   - Why did the president select General Martin to be the next defense minister?
2. Which of the questions below is an example of a question that has a potentially unlimited set of answers that is constrained by the available information?
   - Will the president select General Martin as the country's next defense minister?
   - Which of the three country's four-star generals is the president likely to nominate as the country's next defense minister?

2.2 Types of Hypotheses

Generally, our major hypotheses refer to events or situations that we cannot observe directly. These events may have happened in the past, are now possibly happening, or may occur in the future.

2.2.1 Hypotheses Concerning a Past Event

A terrorist incident occurred two months ago in which several lives were lost. After an investigation, two suspects, X and Y, have been identified. Here are some hypotheses we could entertain about this past event:

H_1: Person X was involved in this incident.
H_2: Person Y was involved in this incident.
H_3: Both X and Y were involved in this incident.
H_4: Neither X nor Y were involved in this incident.

2.2.2 Hypotheses concerning a present event or situation

You might suspect that Country X is still holding prisoners of war taken years ago during a conflict we had with it. Your suspicion here forms one hypothesis, H_5: Country X inow holds someof our POWs. This example illustrates why it is true that we always have more than one hypothesis. Another possibility is $notH_5$, that is, H_6: Country X is not holding any of our POWs.

2.2.3 Hypotheses concerning a future event or situation

We have closely monitored the deteriorating relations between Countries A and B, which share a common border. We now entertain the possibility of armed conflict between these two countries in the near future. Thus, we have a major hypothesis:

> H_7: There will be armed conflict between A and B in the near future.

Another hypothesis, of course, is $notH_7$, That is:

> H_8: There will be no armed conflict between A and B in the near future.

This example allows us to see that we will often need to make our hypotheses more specific. The hypothesis that armed conflict will be between A and B is not very informative if it is our final stated conclusion. Decision makers will wish to know such things as:

- Who will start it?
- How will the conflict proceed? How long will it last? and
- Who will win?

All of these examples concern events/situations that might have happened and are now possibly happening or might happen in the future. We have no certainty about any of these hypotheses. At the moment, they are all simply possibilities. If, at the moment, we reported any of these hypotheses in the form of a conclusion, we would not be taken seriously. We have given no one else any reasons why the hypothesis we have chosen to report as a conclusion should be favored over any of the other hypotheses that are possible. This is where our next two ingredients come in: evidence and arguments.

2.2.4 Hypotheses Generation Strategy

As you may know, Malaysia Airlines Flight 370 disappeared on 8 March 2014 while flying from Malaysia to China. The question is: What happened to it?

Each possible answer to the question is a hypothesis to test. You need to determine which hypothesis is best supported by the available information. Follow the steps indicated below.

Imagine as many potential answers as possible:

Hijacking
Fire
Cyberattack
Shoot down
Alien abduction

Discard any implausible answer (e.g., "Alien abduction"), keeping however those that are high risk but low probability.'

For example, assessing whether a terrorist organization has nuclear weapons, even if assessed to be very unlikely, is still worth assessing because of the enormous implications should the hypothesis be true.

For a binary question, consider first the answer that appears more likely.

Does Hakka have chemical weapons? The available information appears to suggest that it does not, but write the answer as a complete sentence rather than "No:" Hakka does not have chemical weapons.

For a question with a fixed number of answers, because of information gaps, all of the potential answers identified in the problem may be inconsistent with the available information.

For example, if two terrorist groups appear as the prime suspects in a terror attack, but there is information that tends to rule out or minimize the likelihood of their involvement, the correct answer might be an unidentified third terrorist group.

2.2.5 Review Questions

3. A terrorist incident occurred two weeks ago in an American city involving considerable destruction and some loss of lives. After an investigation, two foreign terrorist groups have been identified as possible initiators of this terrorist action: an Al Qaeda-affiliated Group A from Yemen and a Taliban Group B from Pakistan. What are some hypotheses we could entertain about this event?

4. You might have reason to suspect that Iran is now supplying IEDs to a Taliban group in Afghanistan. Since there are other possible sources for these weapons, you will have more than one major hypothesis about possible suppliers of these IEDs. What are some of these other hypotheses?

2.3 Answers

5. Characterize each of the questions below concerning the number of answers it can have.
   a) Will the president select General Martin to be the country's next defense minister?
   b) Which of the country's four-star generals is the president likely to nominate as the country's next defense minister?
   c) Why did the president select General Martin to be the next defense minister?

Answer

a) Binary question because it has only two possible answers, Yes or No.
b) A limited number of answers were given by the number of the four-star generals.
c) Any number of answers, each corresponding to a reason for selecting General Martin to be the next defense minister. The available information will limit this number.

6. Which of the questions below is an example of a question that has a potentially unlimited set of answers that is constrained by the available information?
   a) Will the president select General Martin to be the country's next defense minister?
   b) Which of the three country's four-star generals is the president likely to nominate as the country's next defense minister?

Answer

a) There may be a number of reasons for selecting General Martin as the next defense minister. The available information will limit this number.

b) Thisuestion does not have an unlimited set of answers. The number of answers is limited by the number of the four-start generals.

7. A terrorist incident occurred two weeks ago in an American city involving considerable destruction and some loss of lives. After an investigation, two foreign terrorist groups have been identified as possible initiators of this terrorist action: an Al Qaeda-affiliated Group A from Yemen and a Taliban Group B from Pakistan. What are some hypotheses we could entertain about this event?

Answer

Some hypotheses we could entertain are the following ones:
H1: Group A was involved in this incident.
H2: Group B was involved in this incident.
H3: Both Groups A and B were involved in this incident.
H4: Neither Group A nor B were involved in this incident.

8. You might have reason to suspect that Iran is now supplying improvised explosive devices (IEDs) to a Taliban group in Afghanistan. Since there are other possible sources for these weapons, you will have more than one major hypothesis about possible suppliers of these IEDs. What are some of these other hypotheses?

Answer

- Iran is now supplying improvised explosive devices to a Taliban group in Afghanistan.
- Hezbollah is now supplying improvised explosive devices to a Taliban group in Afghanistan.
- Pakistan is now supplying improvised explosive devices to a Taliban group in Afghanistan.

3 EVIDENCE

3.1 What Is Evidence?

Marcus Aurelius
Everything we hear is an opinion, not a fact. Everything we see is a perspective, not the truth.

Evidence is any observable sign, datum, or item of information that is relevant in deciding whether a statement or hypothesis we are considering (e.g., a scientific or medical claim) is true or false.

Difficulties arise when a variety of terms are used as synonyms for the term evidence: data, items of information, facts, and knowledge. When examined carefully, there are some valid and important distinctions to be made among these terms that are not always apparent, as will we now discuss.

3.1.1 Evidence, Data, and Information

Definition and Examples

Data are un-interpreted signals, raw observations, or measurements. Examples are 137 and STEMQ.

Information is data equipped with meaning provided by a certain context. Examples are: cesium-137, STEMQ Company, and STEMQ customer.

There are untold trillions of data and information items in existence that will almost certainly never become evidence in most inferences. As we will discuss, items of information only become evidence when their relevance is established regarding some matter to be proved or disproved.

Question

9. Provide illustrations of data and information.

3.1.2 Evidence and Fact

Definition and Examples

Now consider the term *fact*; there are some real troubles here regarding its relation to the term evidence. How many times have you heard someone say, "I want all the facts before I draw a conclusion or make a decision?" Or "I want to know the facts in this matter?" The first question is easily answered; we will never have all the facts in any matter of inferential interest. Answers to the second question require a bit of careful thought. Here is an example of what is involved:

Example 1. The locksmith Clyde tells us that the lock of theSTEMQ warehouse had been forced. Now we regard it as fact that Clyde gave us this information. But whether the lock had been forced is only an inference and is not a fact. This is precisely why we need to carefully distinguish between an event and evidence for this event.

Here is what we have: Clyde has given us evidence E*, saying that event E occurred, where E is the event that the lock had been forced. Whether this event E did occur or not is open to question and depends on Clyde's credibility. If we take it as fact that event E did occur, just because Clyde said it did, we would be overlooking the credibility foundation for any inference we might make from his evidence E*. Unfortunately, it so often happens that people regard the events reported in evidence as being facts when they are not. Doing this suppresses all uncertainties we may have about the source's credibility.

We have exactly the same concerns about the credibility of tangible evidence. For example, we have been given an image as evidence E* that we believe reveals the occurrence of event E. But we must consider whether this image is authentic and it is what we believe it to be. In any case, the events recorded in evidence can only be regarded as facts if provided by perfectly credible sources, something we almost never have. As another example, any information we find on the Internet should be considered as only a claim by its source rather than as fact, that is, as evidence about a potential fact rather than a fact.

Question

10. Can you think of an example in which we infer an event E based on evidence E*?

3.1.3 Evidence and Knowledge

Definition and Examples

Now consider the term *knowledge* and its relation with *evidence*. Here is where things get interesting and difficult. As you know, the field of epistemology is the study of knowledge, what we believe it may be, and how we obtain it.

Two questions we would normally ask regarding what Clyde just told us are as follows:

- Does Clyde really know what he just told us, that the lock had been forced?
- Do we ourselves then also know, based on Clyde's testimony, that the lock had been forced?

Let's consider the first question. For over two millennia, some very learned people have been troubled by the question: What do we mean when we say that person A knows that event B occurred? To apply this question to our source, Clyde, let's make an assumption that will simplify answering this question. Let's assume that Clyde is a competent observer in this matter. Suppose we have evidence that Clyde is a professional locksmith, and he actually looked at the lock. This is a major element of Clyde's credibility.

Here is what a standard or conventional account says about whether Clyde *knows* that the car did not stop at the red-light signal.

First, here is a general statement of the standard account of knowledge: *knowledge is justified true belief.*

Person A knows that event B occurred if:

- Event B did occur, [True]
- A got non-defective evidence that B occurred, [Justified]
- A believed this evidence [Belief]

This standard analysis first says that event B must have occurred for A to have knowledge of its occurrence. This is what makes A's belief true. If B did not occur, then A could not know that it occurred.

Second, A's getting non-defective evidence that B occurred is actually where A's competence arises. A could not have gotten any evidence, defective or non-defective if A had not been aware that B could have occurred. Then, A believed the evidence A received about the occurrence of event B, and A was justified in having this belief by obtaining non-defective evidence of B's occurrence.

So, in the case involving Clyde's evidence, Clyde knows that the lock had been forced if:

The lock had been forced,
Clyde got non-defective evidence that the lock had been forced,
Clyde believed this evidence.

If all three things are true, we can state in this standard analysis that Clyde *knows* that the lock has been forced.

Before we proceed, we must acknowledge that this standard analysis has been very controversial in recent years, and some philosophers claim to have found alleged paradoxes and counterexamples associated with it. Other philosophers dispute these claims. Most of the controversy here concerns the justification condition; what does it mean to say that A is justified in believing that B occurred? In any case, we have found this standard analysis very useful as a heuristic in our analyses of the credibility of testimonial evidence.

But now we have several matters to consider in answering the second question: *Do we ourselves also know, based on Clyde's testimony, that the lock had been forced?*

The first and most obvious fact is that we do not know the extent to which any of the three events just described in the standard analysis are true. We cannot get inside Clyde's head to obtain necessary answers about these events. Starting at the bottom, we do not know for sure that Clyde believes what he just told us about the lock being forced. This is a matter of Clyde's veracity or truthfulness. We would not say that Clyde is being truthful if he told us something he did not believe.

Second, we do not know what sensory evidence Clyde obtained, on which to base his belief, and whether he based his belief at all on this evidence. Clyde might have believed that the lock had been forced either because he expected or desired this to be true. This involves Clyde's objectivity as an observer. We would not say that Clyde was objective in this observation if he did not base his belief on the sensory evidence he obtained in his observation.

Finally, even if we believe that Clyde was an objective observer who based his belief about the lock on sensory evidence, we do not know how good this evidence was. Here, we are obliged to consider Clyde's sensory sensitivities and accuracy in the conditions under which Clyde made his observations. Here, we consider such obvious things as Clyde's visual acuity. However, there are many other considerations, such as, "Did Clyde only get a fleeting look at the lock?" For a variety of such reasons, Clyde might simply have been mistaken in his observation; the lock had not been forced.

So, what it comes down to is that the extent of our knowledge about whether the lock had been forced, based on Clyde's evidence, depends on these attributes of Clyde's credibility: his competence, veracity, objectivity, and observational sensitivity.

3.1.4 **Parsing Complex Items of Evidence**

Definition and Examples

Evidence often relates to the occurrence of several events.

<u>Example 2.</u> Here is an item of evidence coming from a human source where the source tells us several things. The source says: I observed person P in company with a known Al-Qaeda operative in Vienna, Austria on August 21. During their conversation, I observed the Al-Qaeda operative taking a stack of $100 bills and a document from Person P. The document looked like a flight manual.

This report, carefully parsed, contains several events:

Person P in company with the Al-Qaeda operative;
The location of their meeting;
The time of their meeting;
The transfer of money and
The transfer of a document looks like a flight manual.

It would be quite incorrect to say that we have just one item of evidence here; in fact we have five items that might be considered separately.

What frequently happens is that a person evaluating evidence such as this may fail to parse the evidence:

- The time and location of the meeting between P and the Al Qaeda operative is one matter.
- What transpired between them is quite another.
- That they met on a certain date and place may bear on one line of argument.
- What transpired between them might bear on another line of argument.

So, it is usually advisable to decompose evidence items into the finest grain details that seem separate to you. Failure to do this can cause lots of further trouble. The reason is that, in our arguments, we want lines of reasoning to be as specific as we can make them.

This same problem can occur with other items of evidence apart from human intelligence. For example, we might encounter the interpretation of a photograph that reveals the occurrence of many events.

We should consider evidence about each event separately so that we will not be confused or confuse others. There are no definite ground rules for parsing complex items of evidence. However, each event being revealed jointly may have a special meaning on its own.

Questions

11. Evidence, especially testimonial evidence, often relates to the occurrence of several events. For example, here is an item of evidence coming from a human source where the source tells us several things. The source says:
    "I observed person P in company with a known Al-Qaeda operative in Vienna, Austria on August 21. During their conversation, I observed the Al-Qaeda operative taking a stack of $100 bills and a document from Person P. The document looked like a flight manual."

This report, carefully parsed, contains several events. Can you identify them?

12. Identify the "dots," details, or "trifles" [as Sherlock Holmes called them] in the following intelligence report:

    FBI Report: [1 April, last Year. Abdul R is the owner of a Gourmet Foods shop in City A, Virginia. [Phone number 703-abc-defg]. BB Union National Bank lists Gourmet Foods as holding account number 10701xxxxxx. Six checks totaling $35,000 have been deposited in this account in the past four months and are recorded as having been drawn on accounts at the Pyramid Bank of Cairo, Egypt, and the Central Bank of Dubai, United Arab Emirates. Both of these banks have just been listed as possible conduits in money laundering schemes.

3.2 Evidence Crdentilas

3.2.1 Relevance

Definition and Examples

The relevance of an item of evidence indicates how strongly this item supports a specific hypothesis in the argument.

Example 3. Suppose your intelligence analysis involves inferences about possible forms of terrorist actions that could be taken against targets here in the United States. Here is an item of information you have just received:

Item #1: Professor Robert Green is the owner of a house in Fairfax County.

The first question you would ask about Item #1 is:

> *So what? What conceivable bearing does this bit of information have on any possible conclusions I could reach about possible terrorist incidents here in the United States?*

Unless Robert Green was associated with any terrorist organization, you would be justified in saying that this datum is totally irrelevant in your present analysis.

The concept of relevance concerns our attempts to answer *"So what?"* questions regarding items of information we have.

Example 4. Now consider a second item:

Item #2: The STEMQ Company in Baltimore, Maryland, manufactures devices for sterilizing medical equipment. A person named Willard reported in a Washington Post article that a canister containing powder, including approximately 3000 curies of cesium-137, had gone missing from the company's warehouse in Baltimore. There were indications that the storage area where this powdered cesium-137 was located showed signs of forcible entry.

Asking the same question, *"So what?"* an analyst would give a different answer:

Cesium-137 has all sorts of uses in medicine and in industry. It is used to sterilize food and manufacture thickness and moisture density gauges, and it is used for various diagnostic purposes in medicine in addition to the sterilization purposes just mentioned. Unfortunately, cesium-137 could be used for other purposes, including the construction of a dirty bomb. Just a few ounces of this powdered material set off by a conventional explosive in a bomb could contaminate an entire city for decades. Indeed, it is known that the Chechen Mujahidin placed a canister of cesium in a Moscow park in the hopes of spreading radiation throughout the city. Fortunately, the police discovered this canister before it was set off. So, Item #2 seems relevant in inferences about what kind of actions terrorist groups may be contemplating here in the United States.

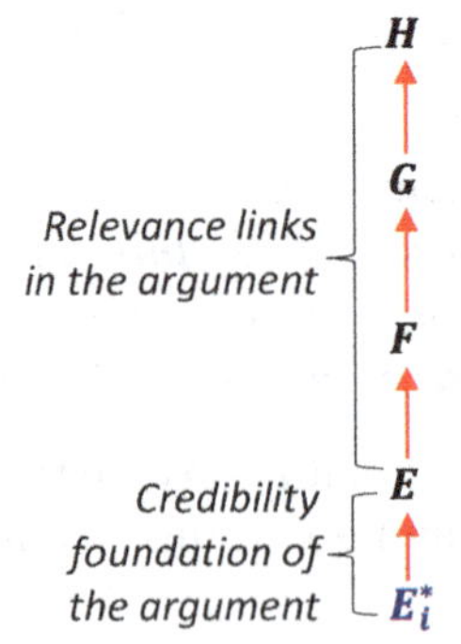

Figure 6. Argument.

The arguments concerning the relevance of evidence involve chains of reasoning or arguments. Consider an item of relevant evidence and the argument or chain of reasoning an analyst has constructed that links this item to the hypotheses being considered (see Figure 6). Here starts the initial root of uncertainty in critical thinking. In forming this chain of reasoning, each identified link in a relevant chain of reasoning involves a proposition that might be true or not, i.e., it is a source of doubt or uncertainty. The argument being constructed thus forms a chain of sources of doubt the analyst believes to be interposed between the evidence and what the analyst is trying to prove or disprove from it. These links or sources of doubt are laid out in a logically consistent order in which one link is inferable from its predecessor.

Where do these links (sources of doubt) come from? They come from the analyst's imagination based on his/her experience and knowledge of the analytic problem area.

Here, we have the roots of disagreements among analysts concerning the relevance of evidence and disagreements about uncertainties in conclusions that may be reached. Different analysts may construct different arguments from the same evidence and thus perceive different sources of uncertainty. Even if they agree about the links in an argument, they may disagree about their strength. It is also true that analysts may generate different possible hypotheses from the same evidence.

A relevant argument has been constructed, and it is now time for its evaluation; here is where the necessity for critical reasoning arises. The analyst must ask the following questions:

- Is the chain of reasoning I have just constructed that shows the relevance of this evidence item logically coherent?
- Does it contain any disconnects or nonsequiturs?

Here is where the necessity for the defensibility of an analytic argument arises. Should anyone else take an analyst's argument seriously as far as the relevance of this evidence is concerned? As we will see in a moment, this question also involves the extent to which anyone else should take the analyst's assessments of uncertainty seriously.

An absolutely crucial element in intelligence analysis is the defensibility of arguments constructed by analysts from the evidence available to possible conclusions to be proved or disproved. Of course, it is also true that an argument must be persuasive and that not all defensible arguments are persuasive. One fairly certain way of failing to make an argument persuasive is to have it revealed that it is not defensible on logical grounds.

But there is an important point here that needs further elaboration:

> *There is no such thing as a uniquely correct or perfect argument from evidence.*

Here is a chain of reasoning an analyst has just constructed to establish the relevance as evidence of an item of information. Someone, another analyst or perhaps another critic, may be able to find what this person believes to be a missing link or a link that is improperly stated. Perhaps this fellow analyst or critic discovers a disconnect in the analyst's initial relevance argument. What no one, be they an analyst or critic, can say is that they have the "correct" argument justifying the relevance of this item of evidence. Someone can come along later and discover inadequacies in the revised argument proposed by the other analyst or critic.

What can be done is to have someone identify defects in an argument. But what no one can do is say that the argument being proposed is the final or ultimately true argument that anyone could propose.

What this says is that someone can correct defects in an argument without ever being able to say that they have the only argument that could ever be made regarding the relevance of evidence. This also accounts for the fact that there will always be some disagreement about the uncertainty that is assessed, combined, and reported among analysts themselves and among persons for whom the analysis was performed. Different persons will construct different arguments from the same bodies of evidence.

The fact that there is no such thing as the ultimately true argument has an important bearing on the many examples we provide of various evidential and inferential issues. We will never say that the arguments we construct in providing examples are the only correct or true ones. You, the reader, may perceive other reasoning routes from the evidence we present and may indeed see other possible conclusions. Our major hope, of course, is that you will not see any disconnects or nonsequiturs in our arguments.

Questions

13. True or false: The relevance of evidence is an assessment of the extent to which the evidence may be believed.

14. Given a hypothesis that a new missile called X was flight tested on 1 July, which of the items of evidence below would have the highest relevance to this hypothesis:

    E1: The missile that was flight-tested demonstrated the same range as missile X.

    E2: A source reported in August that missile X was flight-tested on 1 July.

    E3: A source reported in April that a flight test of missile X was scheduled for 1 July.

15. Given a hypothesis that Mark robbed the bank, which of the items of evidence below would have the lowest relevance to this hypothesis:

    E1: The first five digits of the license plate number of Mark's car match the first five digits of a six-digit license plate number of the car that the bank robber used in his escape

    E2: Mark was not at home at the time of the bank robbery

    E3: Mark told a friend a week before the bank robbery that he was planning to rob a bank

16. True or false:

    Given the hypothesis

    H: John is capable of embezzling money

    the sub-hypothesis

    H1: John was convicted of shoplifting as a teenager

    should be assessed as having lower relevance than

    H2: John has a reputation for having little or no integrity among most of his current co-workers.

3.2.2 **Credibility**

Definition and Examples

Here is an item of evidence E_1^* from source I, that all of us believe is relevant on hypotheses we are considering. We strongly believe that we can defend the relevance of event E, as reported in our evidence E_1^*, by the argument we have constructed that we all agree is very strong and free of disconnects or nonsequiturs. But the crucial question remains:

> *How certain are we that event E did occur just because source I said it did?*

Source I's credibility is the major issue here. Source credibility considerations form the very foundation for all arguments we make from evidence to our hypotheses, as shown in Figure 7. Credentials of evidence. However strong our relevance argument may be, if it rests on a weak credibility foundation, it will falter. Perhaps the best example of inferential calamities known to us in the open-source literature concerns the credibility of the human source called "Curveball" (Bruce, 2008).

The credibility of the source depends on the following credentials: *competence, veracity, objectivity, and observational sensitivity or accuracy.*

Question

17. True or false:
    Source A provides information on subject B. If A is a longstanding enemy of B, the credibility of this information, all other things being equal, should be increased.

3.2.3 Inferential Force or Weight of Evidence

Definition and Examples

In a very general sense, the inferential force or weight of evidence indicates how strong the evidence is in favoring or disfavoring the hypotheses we are considering.

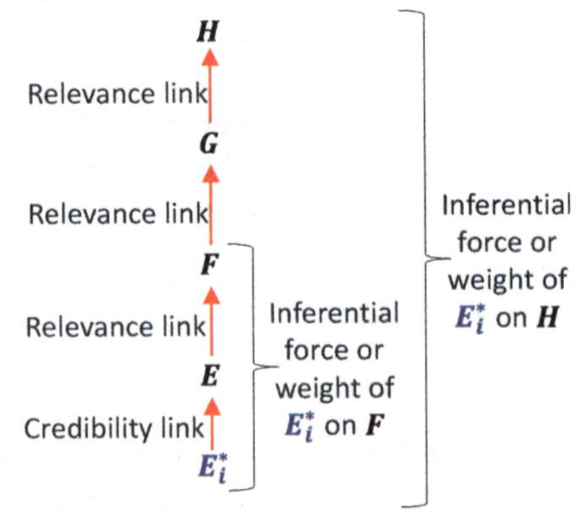

Figure 7. Credentials of evidence.

But this is as far as we can go since there is considerable controversy about what the terms force and weight mean, and especially how these concepts should be assessed and combined. There are only two uncontroversial statements we can make about the force or weight of evidence. First, it has vector-like properties indicating the direction and the strength with which evidence favors or disfavors the hypotheses we are considering. Second, the force or weight of evidence is always graded in probabilistic terms. This second statement is actually the greatest source of controversy since a variety of careful scholars in probability, who have given their days and nights to the study of evidence and probability, cannot agree about how the force or weight of evidence should be assessed and combined.

Probabilistic judgments can be expressed numerically in several ways and also in terms of words. Speaking of numerical judgments of probability, a very wise and devoted scholar, Professor Glenn Shafer, has correctly noted (1988, pp. 5 - 9):

> *"Probability is more about structuring arguments than it is about numbers. All probabilities rest upon arguments. If the arguments are faulty, the probabilities, however determined, will make no sense."*

We add the same concern about verbal assessments of probability, such as very probable, probable, unlikely, and so on. If these arguments are not defensible, no one will take seriously any numerical or verbal assessments we make concerning the force or weight of our evidence.

So, we have found it much easier to provide definitions and meaning to the relevance and credibility credentials of evidence than we have been able to do for the force or weight credential. All we can say at this point is that the force of evidence depends on the strength of our credibility and relevance arguments, as shown in Figure 7. However, as we proceed, we will provide an assortment of examples of how to assess the force or weight of evidence that will be useful in intelligence analysis. Remember that all conclusions in intelligence analysis, in common with all other contexts in which conclusions are based on evidence, must always be hedged probabilistically by some means. This is because, as we will discuss in the next section, our evidence is always incomplete, usually inconclusive, frequently ambiguous, commonly dissonant, and comes to us from sources having any gradation of credibility shy of perfection. As we will observe, alternative views of probabilistic reasoning capture some of these important considerations, but no single view captures them all.

Questions

18. Inferential force is an assessment that takes into account:
    a) the credibility of evidence
    b) the relevance of evidence
    c) both the credibility and relevance of the evidence

19. If the credibility of an item of evidence is low and the relevance of this evidence to a hypothesis is high, the inferential force of this evidence is:
    a) high
    b) medium
    c) low

3.3 **Answers**

9. Provide illustrations of data and information.

Answer

Data are un-interpreted signals, raw observations, or measurements. Examples are: Red, 6.

Information is data equipped with meaning. Examples are:
Red traffic light (color in context has a specific meaning), red tomato, and 6 am

10. Can you think of an example in which we infer an event E based on evidence E*?

Answer

Suppose we have evidence E* from a source named Mouse that event E occurred, where E is the event that Amad M. attended an al Qaeda weapons training class near Madyan in Northwest Pakistan in. Here we had the task of inferring E based on evidence E*.

11. Evidence, especially testimonial evidence, often relates to the occurrence of several events. For example, here is an item

of evidence coming from a human source where the source tells us several things. The source says: "I observed person P in company with a known Al-Qaeda operative in Vienna, Austria, on August 21. During their conversation, I observed the Al-Qaeda operative taking a stack of $100 bills and a document from Person P. The document looked like a flight manual." This report, carefully parsed, contains several events. Can you identify them?

Answer

The report contains the following events:

Person P in company with the Al-Qaeda operative;
The location of their meeting;
The time of their meeting;
The transfer of money and
The transfer of a document looks like a flight manual.

It would be quite incorrect to say that we have just one item of evidence here; in fact we have five items that might be considered separately.

What frequently happens is that a person evaluating evidence such as this may fail to parse the evidence. The time and location of the meeting between P and the Al Qaeda operative is one matter. What transpired between them is quite another. That they met on a certain date and place may bear on one line of argument. What transpired between them might bear on another line of argument. So, it is usually advisable to decompose evidence items into the finest grain details that seem separate to you. Failure to do this can cause lots of further trouble. The reason is that, in our arguments, we want lines of reasoning to be as specific as we can make them. This same problem can occur with other items of evidence apart from human sources. For example, we might encounter the interpretation of a photograph that reveals the occurrence of many events. We should consider evidence about each event separately so that we will not be confused or confuse others. There are no definite ground rules for parsing complex items of evidence. However, each event being revealed jointly may have a special meaning on its own.

12. Identify the "dots," details, or "trifles" [as Sherlock Holmes

called them] in the following intelligence report:

FBI Report: [1 April, this Year. Abdul R is the owner of a Gourmet Foods shop in City A, Virginia. [Phone number 703-abc-defg]. BB Union National Bank lists Gourmet Foods as holding account number 10701xxxxxx. Six checks totaling $35,000 have been deposited in this account in the past four months and are recorded as having been drawn on accounts at the Pyramid Bank of Cairo, Egypt, and the Central Bank of Dubai, United Arab Emirates. Both of these banks have just been listed as possible conduits in money laundering schemes.

Answer

Here are the specific dots or trifles in this report:

- A man named Abdul R. owns the Gourmet Foods shop in City A in Virginia.
- The phone number at Ramazi's shop is 703-abc-defg.
- Ramazi's shop has account number 1070xxxxxxx at the BB National Bank in City A.
- Six checks totaling $35,000, drawn on accounts at the Pyramid Bank of Cairo, Egypt, and the Central Bank of Dubai, UAI, have been deposited in the above account during the past six months.
- The two foreign banks in Egypt and the UAI are possible conduits in money laundering schemes.

An important fact is that any one of these dots or trifles may not seem important at all until they are connected with dots or trifles in other intelligence reports you receive. What we must learn to do is to carefully parse each intelligence report to observe the particular trifles each report reveals. Considered separately or independently, individual dots may have little meaning. Taken together, however, some combination of dots or trifles may allow you to generate some very productive hypotheses as well as new lines of inquiry or questioning.

13. True or false: The relevance of evidence is an assessment of the extent to which the evidence may be believed.

Answer

False. The relevance of evidence is an assessment of the

probability that the hypothesis is true if the evidence is true. The credibility of evidence is an assessment of the extent to which the evidence may be believed.

14. Given the hypothesis
    H: John is capable of embezzling money
    the sub-hypothesis
    H1: John was convicted of shoplifting as a teenager,
    should be assessed as having lower relevance than the sub-hypothesis
    H2: John has a reputation for having little or no integrity among most of his current co-workers.

    Answer

    The sub-hypothesis "John was convicted of shoplifting as a teenager" should be assessed as having lower relevance. The relevance of this sub-hypothesis speaks to John's moral compass as a youth, not his current moral compass. It is possible that John has turned his life around. In contrast, the sub-hypothesis that "John has a reputation of having little or no integrity among most of his current co-workers" speaks to his current moral compass and suggests few of his co-workers trust John to do the right thing.

15. Given a hypothesis that Mark robbed the bank, which of the items of evidence below would have the lowest relevance to this hypothesis:

    E1: The first five digits of the license plate number of Mark's car match the first five digits of a six-digit license plate number of the car that the bank robber used in his escape

    E2: Mark was not at home at the time of the bank robbery

    E3: Mark told a friend a week before the bank robbery that he was planning to rob a bank

Answer

E1 has the highest relevance. It is not certain because someone else may have taken or borrowed Mark's car, and there is a small chance that the sixth digit in the license plate number is different from the license plate number on Mark's car.

E2 proves almost nothing: Mark could have been away from home for any number of possible reasons.

E3 has high relevance but is not conclusive because Mark may have changed his mind; nonetheless, it is much more incriminating than evidence of Mark not being home when the bank was robbed.

Thus, E2 has the lowest relevance.

16. True or false:
    Given the hypothesis
        H: John is capable of embezzling money
    the sub-hypothesis
        H1: John was convicted of shoplifting as a teenager
    should be assessed as having lower relevance than the sub-hypothesis
        H2: John has a reputation for having little or no integrity
    among most of his current co-workers.

17. True or false: Source A provides information on subject B. If source A is a longstanding enemy of subject B, the credibility of this information, all other things being equal, should be increased.

Answer

False: Because A is a longstanding enemy of B, A is biased against B, and A's credibility should be decreased, not increased.

18. Inferential force is an assessment that takes into account:
    - the credibility of evidence
    - the relevance of evidence
    - both the credibility and relevance of the evidence

Answer

Inferential force is an assessment that takes into account both the credibility and relevance of evidence.

19. If the credibility of an item of evidence is low and the relevance of this evidence to a hypothesis is high, the inferential force of this evidence is:

    - high

- medium
- low

Answer

The inferential force is the smallest between the credibility (low) and the relevance (high), and is therefore low.

26. In several accounts of intelligence analysis we have observed the following inference: "We can believe what this source has reported to us because he had good access to the information he reports." Is this inference correct?

Answer

This is a glaring non sequitur. The source may have had all the access in the world, but still not be credible in his report about what he observed. The problem here is that HUMINT asset competence does not entail the asset's credibility; competence is just one of the credentials of credibility.

27. Can you think of other situations in which human competence is an issue?

Answer

As you know, we rely on persons to inform us about the meaning of various forms of tangible and testimonial evidence. In addition, we rely upon many persons who process, edit, and transmit intelligence information of many kinds. Here are some examples:

- Experts who interpret any form of information obtained from sensors;
- Persons who translate documents written in foreign languages;
- Persons who edit, transcribe, or summarize intelligence information;
- Persons who process raw sensor records.

In all of these situations, competence concerns the skills these persons demonstrate when performing their tasks. We might be as misled by a photo that is misinterpreted as we would be if a document written in a foreign language suffered from translation errors.

23. However credible you believe a person might be, you also must give consideration to this person's competence. Provide an example of a credible source which is not believable.

Answer

You might believe that your best friend is certainly a credible source of information. However, you would not consult your best friend regarding a medical problem you face unless your friend is also a physician trained in an appropriate specialty.

20. Our source tells us that event E occurred, but we later find out that E did not occur. Was this source untruthful?

Answer

Not necessarily, since this source might simply have been mistaken, but the veracity of HUMINT sources that report on what they observe depends on our assessment of whether these sources actually believe what they are reporting to us.

21. Can you think of an example of a source that may be subject to any significant exploitation to provide certain information?

Answer

Consider, as an example, a source whose family is detained by Al Qaeda and who has received threats from it to provide us with this information.

22. True or false: Source A provides information on subject B. If A is a longstanding enemy of B, the credibility of this information, all other things being equal, should be increased.

Answer

False: Because A is a longstanding enemy of B, A is biased against B, and A's credibility should be decreased, not increased.

23. However credible you believe a person might be, you also must give consideration to this person's competence. Provide an example of a credible source which is not believable.

Answer

You might believe that your best friend is certainly a credible source of information. However, you would not consult your best friend regarding a medical problem you face unless your friend is also a physician trained in an appropriate specialty.

24. Our source tells us that event E occurred, but we later find out that E did not occur. Was this source untruthful?

Answer

Not necessarily, since this source might simply have been mistaken, but the veracity of HUMINT sources that report on what they observe depends on our assessment of whether these sources actually believe what they are reporting to us.

25. Can you think of an example of a source that may be subject to any significant exploitation to provide certain information?

Answer

Consider a source whose family is detained by Al Qaeda and who has received threats from it to provide us with this information.

4 SOURCE CREDIBILITY

The credibility of the source depends on the following credentials: competence, veracity, objectivity, and observational sensitivity or accuracy.

Albert Einstein
Thinking is hard work; that's why so few do it.

4.1 Competence

4.1.1 Access

The first question to ask related to competence is whether this source actually made the observation he claims to have made or had access to the information that he reports.

4.1.2 Understandability

The second competence question concerns whether this source understood what was being observed well enough to provide us with an intelligible account of what was observed. Thus, besides access, competence also involves understandability.

4.1.3 Questions

26. In several intelligence analysis accounts, we have observed the following inference: "We can believe what this source has reported to us because he had good access to the information he reports." Is this inference correct?
27. Can you think of other situations in which human competence is an issue?
28. However credible you believe a person might be, you also must give consideration to this person's competence. Provide an example of a credible source which is not believable.

4.2 Veracity, Sincerity, or Truthfulness

4.2.1 Definition and Examples

From experience, we learn that people do not always believe what they are telling us. We would not say that a person was being untruthful if this person believed what he/she just reported. So the first question we should ask about the source telling us that event E occurred is:

Does this source believe that event E occurred?
Assessing veracity, sincerity, or truthfulness has been a problem for centuries, and some mistakes have been made in explaining what veracity means. In many old and in some newer works, it is said that a source is being truthful only if the event reported actually occurred. There is great trouble here since this explanation confounds veracity with the two other credibility attributes we need to consider.

4.2.2 **Question**

29. Can you think of an example of a source that may be subject to any significant exploitation to provide certain information?

4.3 **Objectivity**

4.3.1 **Definition and Examples**

From common experience, we observe that persons, including ourselves, often believe that some event has occurred because we either expect it to occur or want it to occur, regardless of what our senses are telling us. In such instances, we would say that this source lacks objectivity. An objective observer is one who bases a belief on the sensory evidence he/she received rather than on what this person expected or desired to observe. Suppose we believe that the source telling us that event E occurred is being truthful; he does believe that event E occurred. But now the question is:

Was this belief based on the sensory evidence this source received, or was it based on what this source expected or wished to observe?
One additional important matter concerns the role of memory. The reason is that our beliefs are elastic in nature; they change over time and often in response to new information we receive. If a HUMINT source made the observation some time ago, we might well question whether this source had the same belief at the time of his/her observation that this person now has while reporting to us. Is this person now telling us what he/she expected or wished to occur instead of basing this report on this person's

recollection of what his/her senses recorded?

Example 5. Suppose we are concerned about the credibility of a source called "Mable." Mable says he observed Yaqub M. placing an IED on a road leading from Baghdad to Samarra two days ago. We have another source, "Foxtrot," who knows "Mable" and who has told us useful things about Mable in the past. So, we tell Foxtrot about Mable's telling us about observing Yaqub M placing the IED on the road between Baghdad and Samarra two days ago. So, we ask Foxtrot if Mable either expected Yaqub or wished Yaqub to be the person he saw placing the IED. Foxtrot says that Mable barely knows Yaqub and has no grounds for expecting or wishing that Yaqub was the person he saw. Foxtrot adds, "You can be confident that Mable told you what he did see and not what he expected or what he wanted to see."

Example 6. Here is Rosebud again, who first tells us that it was Amir D. running away from a car just before a bomb in this car exploded in a crowded market in Baghdad on 14 May 2013, and then two weeks later now tells us that it was Omar T. and not Amir D. This may simply be the result of Rosebud's changing her mind about who she saw, and not the result of her failing to keep her story straight.

4.3.2 Question

30. True or false: Source A provides information on subject B. If source A is a longstanding enemy of subject B, the credibility of this information, all other things being equal, should be increased.

4.4 Observational Sensitivity or Accuracy

4.4.1 Definition and Examples

Suppose we believe our source to be truthful and objective in their report that event E occurred. This source believes that E occurred, and this source bases this belief on sensory evidence obtained during a relevant observation. But now the question is:

How good was the sensory evidence this source received under the conditions in which this observation was made?

As we know, none of our senses are perfectly accurate or

sensitive, particularly under a variety of ambient conditions such as reduced visibility and high noise levels. The physical condition of the source is also relevant here. We would question the adequacy of the sensory evidence this source obtained if he/she had some sensory defect or was intoxicated at the time the observation was made. Common experience tells us that human senses are not infallible and that we are all prone to make mistakes in our observations.

Example 7. Here comes Mable again, who tells us that he observed Yaqub M. placing an IED on a road leading from Baghdad to Samarra two days ago. We ask Mable how sure he is that the person he saw was Yaqub M. Mable says, "Well, I am pretty sure it was Yaqub M. who I saw as I was driving by, but I actually had a brief look at him."

Example 8. Here comes our source, "Dingbat," who tells us that it was Umar who was running away from the car just before a bomb in this car exploded in a crowded market in Baghdad on 14 May 2013. We ask Dingbat how sure he is that it was Umar who was running away from the car. Dingbat says, "I am pretty sure it was Umar, but I can't be sure since it was a foggy day, and I was about half a block away from the car."

4.5 **Answers**

26. In several accounts of intelligence analysis we have observed the following inference: "We can believe what this source has reported to us because he had good access to the information he reports." Is this inference correct?

Answer

This inference is a glaring non sequitur. These accounts all say, "We can believe what this source has reported to us because he had good access to the information he reports." This source may have had all the access in the world, but still not be credible in his report about what he observed. The problem here is that HUMINT asset competence does not entail the asset's credibility; competence is just one of the credentials of credibility.

27. Can you think of other situations in which human

competence is an issue?

Answer

There are other situations that are important in intelligence analysis in which human competence is an issue. As you know, we rely on persons to inform us about the meaning of various forms of tangible and testimonial evidence. In addition, we rely upon many persons who process, edit, and transmit intelligence information of many kinds. Here are some examples:

a) Analysts who interpret any form of information obtained from sensors;
b) Persons who translate documents written in foreign languages;
c) Persons who edit, transcribe, or summarize intelligence information;
d) Persons who process raw sensor records.

In all of these situations, competence concerns the skills these persons demonstrate when performing their tasks. We might be as misled by a photo that is misinterpreted as we would be if a document written in a foreign language suffered from translation errors.

28. However credible you believe a person might be, you also must consider this person's competence. Provide an example of a credible source which is not believable.

Answer

You might believe that your best friend is certainly a credible source of information. However, you would not consult your best friend regarding a medical problem you face unless your friend is also a physician trained in an appropriate specialty.

29. Can you think of an example of a source that may be subject to any significant exploitation to provide certain information?

Answer

Consider, as an example, a source whose family is detained by Al Qaeda and who has received threats from it to provide us with this information.

30. True or false: Source A provides information on subject B. If source A is a longstanding enemy of subject B, the credibility of this information, all other things being equal, should be increased.

Answer

False: Because A is a longstanding enemy of B, A is biased against B, and A's credibility should be decreased, not increased.

5 METHODS OF ASSESSING UNCERTAINTY

5.1 Major Sources of Uncertainty in Masses of Evidence

There are five major reasons why conclusions reached in intelligence analysis based on evidence will be necessarily probabilistic in nature: our evidence is always incomplete (we can look for more if we have time), usually inconclusive (it is consistent with the truth of more than one hypothesis or possible explanation), frequently ambiguous (we cannot always determine exactly what the evidence is telling us), commonly dissonant (some of it favors one hypothesis or possible explanation but other evidence favors other hypotheses), and has various degrees of credibility. Any one of these reasons can lead to uncertain conclusions, but an analyst drawing conclusions based on masses of different forms and combinations of evidence will likely encounter all five of these reasons at the same time. Intelligence analysts who report conclusions with varying degrees of uncertainty are often unjustly criticized for doing so. Persons providing such criticisms are doubtless unaware of many or most of the reasons why it is necessary to hedge conclusions in probabilistic terms.

Pierre-Simon Laplace
Probability theory is nothing but common sense reduced to calculation.

One main reason we have for providing a careful account of these five reasons is that no single view of probability we know about captures all five of these sources of uncertainty well. Each view of probability we will mention provides useful insights about some of these sources of uncertainty, but no single view says all there is to be said.

5.1.1 Incompleteness

We may have all heard someone say, "I am going to wait until I have all the evidence before I draw a conclusion or make a decision." This person faces an infinitely long wait because there is no situation in which we can say we have all the possible evidence. The first way of showing that this statement is true is to consider the distinction we have made between directly relevant and ancillary evidence. In doing so, we also referred to ancillary evidence as being meta-evidence or evidence about evidence. The trouble here is that we face an infinite regress in which we have evidence, ancillary evidence about this evidence, ancillary evidence about this ancillary evidence, and so on, ad infinitum. Suppose we have a human asset that provides us with some interesting evidence; call this asset our primary source. However, we have a secondary source that provides us with ancillary evidence about the credibility of our primary source. But then we have a tertiary source that provides ancillary evidence about the credibility of our secondary source. This process could go on and on indefinitely. This fact was noted years ago by the CIA's James J. Angleton, who encountered situations in which chains of human sources provided contradictory and divergent evidence about each other's credibility. He described this situation as similar to being in a "wilderness of mirrors" (Martin, 1980).

There are many other situations in which we could have endless chains of evidence and evidence about evidence, even in empirical statistical situations. It is often said that the conclusions reached by statisticians are always misleading to some extent and that they must choose ways of reporting conclusions that are minimally misleading. Here is an intelligence analyst who reports the results of a statistical analysis of the capabilities of a weapon system of some sort. Her conclusions are challenged by another analyst. In turn, the comments made by this challenger are then challenged by a third analyst. This process could go on forever until someone decides to call a halt to this evidence about evidence situation. This fact is noted in our procedures for trials at law that limit the extent to which we can follow this chain. If this were not the case, a trial might go on for years without any verdict ever being reached.

However, there is another even more important reason for our evidence always being incomplete. In how many intelligence analyses could it ever be said that every one of the relevant questions that could have been asked was, in fact, answered by the evidence that was gathered and analyzed? There probably has never been an intelligence analysis in which there were no lingering unanswered questions at the time a conclusion was required. In the absence of clairvoyance, there may be considerable uncertainty about what questions should be asked in an intelligence analysis. On various matters concerning events in the past, it might be argued that we have complete evidence, as briefly discussed below.

Example 9. We believe we have evidence that allows us now to conclude, beyond all shadow of doubt, that the twin towers of the World Trade Center in New York City were destroyed on September 11, 2001, and that the then President of the United States, John F. Kennedy, was assassinated in Dallas, Texas, on November 22, 1963. We can go to New York and see for ourselves where the World Trade Center used to be and then view television images of the two aircraft that slammed into the buildings and the horror of the subsequent collapse of the buildings. We can also view the monument to John F. Kennedy in Arlington National Cemetery in Virginia, which records the place and time of his death. We can also watch the so-called Zapruder film that shows President Kennedy being struck in the head and then collapsing into his wife Jacqueline's arms after he was struck. In each of these cases, we have what can be regarded as conclusive evidence.

There are two points to be made about the two examples we have just provided. First, it does appear that the occurrence of these two events is no longer a topic of analytic interest since we are certain that they occurred. However, there are many other questions concerning these two events that are now of great interest and will continue to be of interest in future analyses. Concerning the tragedy in New York, we are asking such questions as, "Who were all the persons involved in the planning of this action?" and "Why did these persons take this action at this particular time?" As far as the Kennedy assassination is concerned, it is still being asked, "Did Lee Harvey Oswald act by himself, or was he part of a conspiracy to kill President Kennedy?" The evidence on these questions will never be complete. The second point concerns the stability of the conclusions we have now reached with certainty over time. Five hundred years from now, will these two conclusions about the destruction of the World Trade Center and the assassination of President Kennedy still be regarded as certain (if they are, indeed, matters of concern at all)? Perhaps the evidence we now regard as conclusive will have vanished long before the year 2514.

Intelligence analysts will not normally be concerned about what will be inferred five hundred years from now about events of interest today. In many cases, they are asked to reach conclusions concerning past events based on evidence that is far from complete. For example, new accounts of the identities and motivations of the nineteen terrorists who destroyed the World Trade Center appear regularly. Now, consider an intelligence analysis that involves the prediction of some future event, such as whether countries such as Iran, Iraq, Saudi Arabia, Syria, Egypt, and Turkey will engage in an all-out war if we remove all of our forces from countries in the Middle East. Unless we have a person who is certifiably clairvoyant and can see into the future, we will never know for certain. The only thing certain is that we can only draw inferences about events such as these based on evidence about events in the past and the ever-fleeing present. This evidence will obviously be incomplete, if only because we may not be asking the right questions. In addition, and of the greatest importance, there will always be questions unanswered by the evidence we do have. There is only one view of probabilistic reasoning, the Baconian view (Cohen, 1973; Tecuci and Schum, 2023), that asks how complete is our coverage of evidence and how many questions remain unanswered by the evidence we have considered.

5.1.2 **Question**

31. We suspect that person P is a double agent and is presently passing classified information to a certain potential adversary. We thought he worked only for us but now have some grounds for a belief that he is also employed by this adversary. What would constitute absolutely complete evidence that P is in fact a double agent?

5.1.3 **Inconclusiveness**

The evidence encountered in intelligence analysis is commonly inconclusive, meaning that it is consistent with more than one possibility, hypothesis, or explanation. Conclusions reached from such evidence can only be probabilistic in nature. Conclusive evidence, which is consistent with only one possibility or hypothesis, is usually in very short supply. Conclusive evidence would supply complete grounds for, or make necessary or certain, some hypothesis or possible conclusion.

Intelligence analysts use the term nugget to refer to conclusive evidence on some major hypothesis that comes from a completely credible source.

Example 10. Here is an example of such a nugget. Suppose we have a trusted source who reports the following events: "Persons associated with the North American Jihadist Organization (NAJO) in Silver Spring, Maryland, acquired the cesium-137 that was stolen from the STEMQ warehouse in Baltimore. They are now constructing a dirty bomb in the garage of a residence at 221 Colesville Rd. in Silver Spring, Maryland, which they intend to set off on the grounds of the capitol building in Washington, DC, next Thursday at 12:00 PM."

Example 11. If we had such a nugget, we could easily prevent this disaster from occurring. Barring the acquisition of such a nugget we must, as the phrase goes, "mine lots of lower grade ore" in the form of inconclusive evidence.

5.1.4 **Question**

32. One of our military transport aircraft made a stop at a civilian airfield in Country C, with whom we have had friendly relations over the years. Two days ago, this aircraft was destroyed on the ground by an explosive device. There are identified groups in Country C that do not favor C's continued friendly associations with us; one of these groups is called the "Purples." Person Q, believed to be associated with the "Purples," was observed by a usually believable source in an aircraft parking area of this airfield just one

hour before our aircraft was destroyed. Why is this just inconclusive evidence that the "Purples" were involved in this incident?

5.1.5 **Ambiguity**

Evidence is ambiguous to the extent to which we cannot determine what it is telling us. We might also describe ambiguous evidence as being imprecisely stated. In addition, ambiguous evidence goes beyond being merely inconclusive. We may have precisely stated, non-ambiguous evidence that is still consistent with more than one hypothesis or possible conclusion.

<u>Example 12.</u> As an example of ambiguity, consider a conversation we have intercepted between two persons of interest who are involved in known terrorist activities. In this conversation, they make use of words to describe persons and their activities that are designed to conceal their intentions or to mislead others who may be listening. It is well known, for example, that terrorist organizations in the Middle East spend a great amount of time training their operatives to be skillful in disguising their identities, capabilities, and intentions. Even our best efforts to disambiguate the evidence we receive are not always successful. A person listening to such a conversation may be perfectly fluent in the language and dialect being used by the two persons overheard but still be unable to tell us exactly what these two persons were saying.

<u>Example 13.</u> In many cases, we may encounter information that would seem relevant if we could decide what it is telling us; in other words, it is ambiguous, imprecise, or vague. We may, for example, be provided with a document concerning apparently current plans of a certain terrorist organization. This document, on careful examination, is reckoned to be authentic. The document says, "Our destiny is now approaching. We will meet at the usual place at the agreed-upon time and proceed to the Crusader site. If Allah is willing, we will kill many of them." This message is certainly vague with respect to the time and place of what seems to be an intended terrorist operation. With any luck, we might have other evidence about this group that

might allow us to remove at least some of the ambiguity apparent in this document. The basic trouble with ambiguous evidence is that it certainly generates uncertainty, but we are often at a loss about how to remove at least some of this uncertainty.

Example 14. In many cases involving intelligence reports you receive, you cannot tell exactly what information is being conveyed in these reports. Very good examples are provided by the equivocal testimonial evidence. Here is a human source who hedges when asked if event E occurred. The source says, "I am not sure, but I believe it is about 60% likely that event E happened." The trouble is that "60%" for this source might not be the same as what you believe "60%" means. Evidence is frequently encountered that is imprecise in some way, resulting in the inability to tell exactly what event is being reported. If you can't be sure what your evidence says, you can hardly avoid hedging your conclusions.

Example 15. In so many instances, reports the sources provide ambiguous accounts of observed events. Here is a source who reports seeing "A large number of Taliban fighters assembling, a short time ago, near location X in Afghanistan." This is an imprecise or fuzzy account of this situation. Hearing this account, we do not know exactly how many, where, and when this assemblage of Taliban fighters occurred. Was the exact number 20, 50, 100, 500, or larger? Was the time 12 hours ago, 3 hours, 30 minutes, or shorter? Was the location 10 km, 5 km, 2 km, or only 500 m from location X? Why do such ambiguous reports so often occur? The reason is quite simple. Human observers cannot make precise judgments under many conditions and are trying to do their best to honestly report what they observe. Suppose, instead, this source had given the following report: "At exactly two hours and 17 minutes ago, I observed exactly 257 Taliban fighters assembling exactly 0.76 km from location X." We might be quite suspicious of a source who provided such a precise account, especially when we hear about the conditions under which the source said he made this observation.

Example 16. Often, analysts are criticized for providing

ambiguous conclusions in their analyses. For example, an intelligence analyst says her analysis shows that it is quite probable that the Russians are increasing the number of their air defense forces in Damascus, Syria. She is criticized for not providing a precise probability to go along with her conclusion. She says, "OK, you want me to give you a precise numerical probability here, and so I will say that the probability is exactly 0.73 that the Russians are increasing the number of their air defense forces in Damascus in Syria." After she provides this number, she tells a colleague: "I have just responded to a stupid request. I was asked to give a precise probability of the Russians increasing the number of their air defense forces in Damascus, Syria. I said this probability was 0.73. But I certainly did not base this number on 100 past occasions on which the Russians increased the number of their air defense forces 73 times. No one has such a record. The number I gave is just a good guess about what, for me, 'quite probable' means. But this number might actually be any number between 0.65 and 0.85."

5.1.6 Questions

33. We believe the person shown in the photograph is P. However, the figure is blurred. Why is this evidence ambiguous?
34. A radar image shows the possibility of one or more aircraft at a certain location. Why is this evidence ambiguous?
35. An observer tells us that he saw a tall man with very dark hair driving away in an old car from the sight of a terrorist incident shortly after it occurred. Why is this evidence ambiguous?
36. From your own experience, can you recall other items of ambiguous evidence you have received?

5.1.7 Dissonance

Suppose we have an analysis in which it is stated that all of the evidence analyzed was harmonious or directionally consistent in favoring the conclusion reported in the analysis. The first question someone should ask is, "Are you sure you did not gather and analyze just the evidence you believe would favor the conclusion you reported?" In some situations, this is called "Cherry picking." At least some pattern of dissonance may be expected in any intelligence analysis, especially one that involves any degree of complexity. To be sure, some of the evidence may be harmonious in pointing in one direction, but on balance, we will have other harmonious evidence pointing in another direction. In short, some dissonance will be expected in every intelligence analysis having any degree of complexity. Failure to report this dissonance will arouse justified suspicions of the intentions and the competence of the persons performing the analysis.

5.1.8 **Imperfect Credibility**

Sources of evidence of any kind have any possible degree of credibility. No mechanical or electronic sensor is perfectly accurate, and on occasion, sensor records of various sorts can even be faked. As we all know, no human observer is always perfectly credible. In some cases, a person may not even be a competent source of information. Here are some examples of these distinctions.

<u>Example 17.</u> Radars and other sensory images are not always sensitive enough to discriminate between events of interest to us. On occasion, we can obtain estimates of a sensor's hit rate and false-positive rate. It is certainly not unheard of in the intelligence business for documents, photographs, and other similar items of evidence to be faked or forged. At least some items of evidence you may encounter will not be authentic; they are not what they seem to be. Human sources are not always truthful, objective, or accurate as observers. Further, a person may be truthful, objective, or accurate about some matters but not about others. Just because a human source has given us what we take to be believable evidence in the past is no guarantee that he will continue to do so in the future. Failure to recognize this simple fact has produced more than one intelligence-related catastrophe.

An experienced analyst reading the account of uncertainty in intelligence evidence just provided may easily recognize all five of these characteristics and will have encountered most of them at one time or another. Any critic of intelligence analysts for expressing their uncertainty in reporting conclusions should be made aware of the five reasons we have mentioned that require intelligence analysts to acknowledge the extent of their uncertainty, which will always be present. However, there are additional reasons why uncertainty is evident in all complex activities, such as intelligence analysis.

5.1.9 Question

37. The competence, veracity, objectivity, and observational sensitivity need to be considered for all of the human sources in our dirty bomb example. Pick a human source in this example and state what kinds of questions you would ask about the competence, veracity, objectivity, and observational sensitivity of the person you have chosen.

5.2 General Classes of Probability and Uncertainty

5.2.1 Definitions and examples

We have now considered the major sources of uncertainty related to evidence and have shown how uncertainty arises in chains of reasoning linking evidence to hypotheses we entertain. A major credential of evidence we have only mentioned briefly is its inferential force or weight. As we noted, this credential is always expressed in probabilistic terms but is a source of great controversy. In Figure 8, we showed how the force or weight of evidence concerns the strength of all the credibility and relevance links in our chains of reasoning. However, this illustration concerns the chains of reasoning from one item of evidence. In any intelligence analysis, there will be many items of evidence to consider, and very many sources of uncertainty or doubt will be associated with complex arguments that link this mass of evidence to the hypotheses at issue in the analysis. It would not be uncommon to be able to identify hundreds or even thousands of sources of doubt arising from masses of evidence being considered.

There is little consensus about how uncertainty should be expressed, combined, and reported. Most analysts have learned in school about the conventional system of probability in which uncertainty is expressed by percentages, or odds and odds ratios, as the only way in which uncertainty can be expressed about evidence-based conclusions. Many analysts will have taken courses in statistics in which the probability of events is estimated based on the relative frequency of their observed occurrence in the past. This approach involves events that are the result of replicable or repeatable processes that can be counted, as in the following instances.

Example 18. Counts can be made of the frequency of occurrence of certain types of vehicles entering or leaving a certain military installation that is under observation. Voting patterns are examined in a certain country to see how strongly, in the past, voters in a certain country have supported candidates whose interests seem favorable to our own interests. Counts are made of the number of instances of car bomb attacks in Iraq or in other places.

Example 19. There are thus many instances in intelligence analysis in which statistical estimates of probabilities can be made. However, the problem is that there are many more instances in intelligence analysis in which we have uncertainty about certain past or future events when we have nothing to count because these events are unique, singular, or one-of-a-kind. If they happened in the past, they did so on just one occasion. If they will occur in the future, they will do so on just one occasion. In such instances, we will have no statistics to back up our uncertainty assessments. Examples of uncertainty about these unique or singular events abound in intelligence analysis.

Example 20. Did the Iranians supply the shaped explosive devices that destroyed the two Humvees in Iraq? Was this informant or asset truthful or accurate when he identified the person who drove the truck carrying the bomb that destroyed the hotel in Baghdad? Did the Saudis supply the weapons found in the possession of the Sunni militia group? Who will be the successor to Putin in assuming the leadership in Russia?

Example 21. Because there is nothing to count and, therefore, no statistics to support the probabilities associated with answering such questions, the assessments must be judgmental or subjective in nature. This also means that different analysts may assess these probabilities differently and arrive at different probabilities regarding major answers.

The point here is very simple: All statistical reasoning is probabilistic in nature, but not all probabilistic reasoning is statistical in nature.

There are some very interesting but difficult issues concerning the extent to which the concepts and methods so useful in statistical analyses continue to apply in situations in which we have uncertainty but no statistics. There are many issues regarding the assessment, combination, and reporting of uncertainty in these non-statistical situations that are very important but frequently go unrecognized.

Accurately representing uncertainty is not a challenge unique to intelligence analysis. Conclusions reached in law, medicine, science, and history typically are qualified or hedged to convey the extent to which a conclusion falls short of being certain. It is common in any of these professions to qualify one's conclusion (Twining, 2003).

Table 2 categorizes some alternative views of probability. We refer to the views in the first column as enumerative because they assume that the probabilities are the result of counting. We refer to the views in the second column as epistemic since they assume that probabilities are based on some kind of knowledge, whatever form it may take. In short, probabilities are the result of informed judgments.

Table 2. Alternative views of probability.

Enumerative	Epistemic
Aleatory (Chances)	Subjective Bayesian
Relative Frequency	Belief Functions
and Statistics	Baconian Probability
Bayesian Statistics	Fuzzy Probability

5.2.2 Question

38. Provide an example showing how an analyst's numerical assessment of a probability applied to a conclusion can invite criticism.

5.3 Complementariness of Probability Views

Questions about uncertainty are naturally linked to views about probability. Some analysts may have studied probability quite extensively; many others will have little or no formal tutoring on the subject. Still, others may dislike the sometimes complicated formulas that the study of probability often requires. However, words and pictures - rather than mathematics - can express very useful ideas about probability and uncertainty. One does not need a background in mathematics or statistics to reason with uncertainty and draw conclusions from evidence.

Example 22. Consider the scenario where three competing hypotheses on an adversary, H1, H2, and H3, were analyzed for a battlefield commander. A body of evidence was examined and the Bayesian, Belief Functions, and Fuzzy methods have been employed. Each method showed that, based on the existing evidence, H3 has the highest probability (and very close to certainty):

P(H3) = 0.99 Belief(H3) = 0.989 Fuzzy(H3) = almost certain

As time passes and action is taken based on H3, it becomes painfully clear to the battlefield commander that H1 was the true hypothesis, not H3. What went wrong - after all, Subjective Bayes, Belief Functions, and Fuzzy is each a highly respected probability view. Then, an analyst with knowledge of the Baconian probability view enters the debate and makes the following comments:

It is true that your analysis rested on quite a bit of evidence. But how many relevant questions can you think of that were not answered by the evidence you had? If you believed that these unanswered questions would supply evidence that also favored H3, you were misleading yourself since you did not obtain any answers to them. It is now clear that the answers to these questions did not favor H3 but favored H1. So, the posterior probability you determined, by itself, is not a good indicator of the weight of evidence. What makes better sense is to say that the weight of evidence depends on the amount of favorable evidence you have and on how completely it covers matters you said were relevant. In your analysis, you completely overlooked the inferential importance of the questions your existing evidence did not answer.

The Bayesian, Belief Functions, and Fuzzy probability views all answer the question: *How strong is the evidence we do have about this hypothesis?*

However, the Baconian view answers a different question: *How much evidence do we have about this hypothesis, and how many questions about it remain unanswered?*

Table 3. Complementary probability views.

Evidence Characteristic	Subjective Bayes	Belief Functions	Baconian	Fuzzy
Incompleteness			☑	
Inconclusiveness	☑	☑	☑	☑
Ambiguity		☑		☑
Dissonance	☑	☑	☑	☑
Credibility	☑		☑	

Answering the Baconian question would have revealed that there are unanswered questions whose answers may make H3 less likely. Thus, it would have helped to acknowledge that the answers to these unanswered questions may not favor H3.

Notice that, as summarized in Table 3, neither the Subjective Bayesian, nor the Belief Functions, or the Fuzzy probability view can account for the incompleteness of the coverage of evidence. The only view that can account for this incompleteness is the Baconian view (Schum, 2001b; Tecuci et al., 2016a).

What is especially of interest to us is the fact that both the Baconian system and the Fuzzy system use the min/max rules for combining probabilistic assessments:

P(H1 and H2) = minimum{ P(H1), P(H2)}

P(H1 or H2) = maximum{ P(H1), P(H2)}

This gave us the liense to define a combined Baconian and Fuzzy probability system that can cope with all the five characteridtics (see Section 5.5).

5.4 Decomposed versus Holistic Analyses

There are alternative ways of describing the complexity of the tasks we face in the analyses based on masses of different forms of evidence coming from different sources. One way, presently very common, is to say that we have very difficult tasks of "connecting the dots," where we have masses of dots to be connected, and these dots can be connected in various complex ways. Another way is to describe our inferences about major hypotheses as being exercises in drawing final conclusions about matters of interest involving many sources of doubt that arise from the evidence we are considering and the arguments we have constructed based on this evidence. Here come two questions of vital importance in such complex evidential reasoning tasks:

How many sources of doubt should we try to identify and combine?

How many dots should we try to capture and connect?

These questions have a virtually unlimited number of possible answers depending on our time for analysis, objectives, and resources. These same important questions arise in many contexts besides intelligence analysis, such as in law, medicine, science, and many other areas of interest.

We start by discussing two extremes in answers to the above questions. The first involves the Wigmorean analysis (Wigmore, 1913; 1937; Schum, 1987; 1994). We owe a great debt to the field of law for the very rich legacy of experience and scholarship it has provided concerning the credentials of evidence and our drawing conclusions from it. As far as our present purposes are concerned, no one has contributed more to this rich legacy than an American evidence scholar named John H. Wigmore [1863 - 1943]. Wigmore's views on evidence are still widely cited in current discussions of evidence in the field of law. Wigmore was the very first person to study the construction of chains of reasoning in arguments such as those we have been considering. He was the first person known to us to study systematically the task of drawing conclusions from masses of different kinds of evidence. Today, we refer to such a process as involving inference networks. Wimorean networks were further developed by David Schum, William Twinning, and Terrence Anderson.

The intelligence analyst is quite accustomed to using expert judgment as a basis for analyses delivered to senior U.S. policymakers. The analyst might cull the judgments of several other analysts to integrate several lines of investigation and analysis into a complex assessment. In any case, suppose that the analyst has generated and considered several possible conclusions and has decided that the relevant and credible evidence most strongly favors a particular conclusion. Presumably, the analyst reached this conclusion by considering many sources of doubt or uncertainty about how strongly the evidence favored a particular conclusion. But now the question arises: How certain can the analyst be that the chosen conclusion

Figure 8. The forefathers of Wigmorean analysis.

reached is the correct one, and how can uncertainty about this conclusion be communicated to supervisors and policymakers? Failing to accurately report uncertainty about a conclusion risks misleading customers, which could have grave consequences for U.S. foreign policy.

The Wigmorean approach encourages us to try to identify all the sources of doubt or dots we can imagine could affect the conclusions we reach based on the evidence we can defend as being relevant to these conclusions. This approach encourages us to decompose our evidential reasoning task to as detailed a level as we can. In other words, the Wigmorean approach encourages us to drill down to as deep a level as we can.

At the other extreme, we could entertain an extreme level of what is called holistic analysis. Holistic analyses involve various levels of suppression of doubts we may either ignore or don't even attempt to identify in forming our final conclusions. We do everything in our heads, according to methods we can never articulate, in expressions of our uncertainty about a stated conclusion.

Example 23. Here is the most extreme level of holistic analysis possible in answering the following question:

How will the autocratic leader L in Country A respond to demonstrations in several major cities protesting rising bread prices and calling for the ouster of the leader?

In this extreme form of holistic analysis, we don't drill down at all. Here is the ultimate hypothesis we are considering:

H: The leader L will brutally repress the protests, using loyal police and military forces, arrest thousands of protesters, and execute hundreds as foreign agents accused of instigating the riots.

An analyst, or group of analysts, considers this hypothesis and evidence bearing on it and says: "I [or we] have considered the evidence for this hypothesis and can conclude that it is very likely to be true. If you require numbers, I can say that it has a probability of 0.7 of being true". That's all; no further words are provided to indicate how the strength of this conclusion was determined. Asked by a "customer" to state the basis for this fuzzy or exact probability, the analyst might experience varying levels of difficulty in coming up with answers to this question. What the analyst is doing here is saying: "Take my word for it, H is very likely, or having 0.7 probability, of being true". This would hardly be acceptable to most decision-making consumers of intelligence analyses.

Here is a slightly less extreme form of holistic analysis and one that drills down to only a very shallow level of detail. Suppose the analyst, or analysts, says: "I have considered various arguments resulting from the parsing of our major hypothesis *H*. They are as follows:

H1: The leader will brutally repress the protests, using loyal police and military forces.

H2: The leader will arrest thousands of protesters.

H3: The leader will execute hundreds as foreign agents accused of instigating the riots.

The analyst continues: "Based on the evidence we have so far, I think H1 is extremely probable (0.95); H2 is very probable (0.75); and H3 is only fairly probable (0.60). On this basis, I conclude that hypothesis H is fairly probable, say 0.60". In this case, the analyst has only given the barest description of the

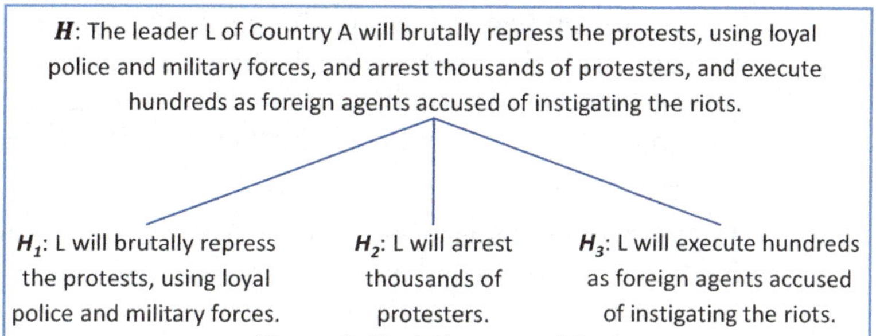

Figure 9. Task decomposition.

probabilistic ingredients of his/her final hedge on hypothesis H. However, the analyst has given no further information about the basis for these four probabilities or how they were combined. This level of analysis would very likely be no more appealing to a decision-making consumer than the most extreme holistic analysis mentioned above.

Any number of further levels of task decomposition or drilling-down are possible. For example, we could marshal our available evidence to see what lines of argument we could generate on each one of the three major lines of argument we have constructed so far on our hypothesis H. This would result in the identification of more sources of doubt. However, our arguments begin to be the most complex inference networks, and the largest number of sources of doubt will be identified as we attempt to link specific patterns or combinations of our evidence to each one of these major arguments. The deeper we drill down or decompose our analysis of complex evidential reasoning tasks we face, the more sources of doubt or uncertainty we will identify, and the more difficult the probability assessment and combination tasks we will face, regardless of the probability system we employ. A basic fact about this complex activity is that no probabilistic or other formal theory associated with the analysis of inference networks will tell us what our constructed networks should be and how many sources of doubt we must identify for any particular complex evidential reasoning task we face.

The construction of an argument based on evidence is always grounded in imaginative reasoning. However, there is another vital form of reasoning involved in the construction of arguments: critical reasoning. We must subject our arguments to very critical analyses in which we attempt to find any disconnects or nonsequiturs in them. So, careful argument construction involves both imaginative and critical reasoning on the part of analysts.

Regarding the persuasiveness of our analyses and the arguments upon which they rest, it seems easy to say that if we have reached imaginative and productive conclusions, resting upon defensible arguments, and if our conclusions are hedged by some appropriate probabilistic method, then our analysis will be persuasive. The trouble, of course, depends upon the people we are trying to persuade. Known since the time of the ancient Greeks, some persons fail to be persuaded by valid arguments and often are persuaded by invalid arguments. As we all learn, the "customers" of our intelligence services may form their own beliefs about matters of vital interest to our nation's security and may disregard even the most imaginative, defensible, and productive intelligence analyses. This is an argument that the "customers" of our intelligence services should also receive the same kind of training that you are now receiving in this book.

5.5 Answers

31. We suspect that person P is a double agent and is presently passing classified information to a certain potential adversary. We thought he worked only for us but now have some grounds for a belief that he is also employed by this adversary. What would constitute absolutely complete evidence that P is, in fact, a double agent?

Answer

Suppose we begin to list all of the factors we should consider before we can say that we are able to give a complete account of P's presently being a double agent. This list grows very long since we would need a complete account of all of P's activities in the past and a complete account of all associations he might have had with any potential adversary. Even if we were willing to say that we had listed all relevant factors we should consider, we would certainly not be able to obtain evidence bearing on all of these factors. For example, it seems very unlikely that any potential adversary would willingly provide us with information bearing on any association they have had with P.

32. One of our military transport aircraft made a stop at a civilian airfield in Country C, with whom we have had friendly relations over the years. Two days ago, this aircraft was destroyed on the ground by an explosive device. There are identified groups in Country C that do not favor C's continued friendly associations with us; one of these groups is called the "Purples." Person Q, believed to be associated with the "Purples," was observed by a usually believable source in an aircraft parking area of this airfield just one hour before our aircraft was destroyed. Why is this just inconclusive evidence that the "Purples" were involved in this incident?

Answer

Evidence is inconclusive to the extent that it could be plausibly explained in more than one way. Here are some plausible reasons why this evidence about Q is only inconclusive evidence that the "Purples" were involved in this incident:

- Q might not, in fact, have any association with the "Purples;" we only believe he has;
- Q might have been in the aircraft parking area in order to board an airplane himself and not with any mischief in mind;
- It might not have been Q that our source said he observed; all we have said is that this source is "usually" believable.

All of these factors, plus others that could be identified, make this evidence about Q only inconclusive in nature. Although this evidence does provide some grounds for a belief that the "Purples" were involved in this incident, it is certainly explainable on other grounds as well.

33. We believe the person shown in the photograph is P. However, the figure is blurred. Why is this evidence ambiguous?

Answer

Because the figure is blurred, we cannot be sure that it is P who is in the photograph. Therefore this evidence is ambiguous.

34. A radar image shows the possibility of one or more aircraft at a certain location. Why is this evidence ambiguous?

Answer

This evidence is ambiguous because we cannot tell how many aircraft are at this location. There might be just one aircraft, but there might be several.

35. An observer tells us that he saw a tall man with very dark hair driving away in an old car from the sight of a terrorist incident shortly after it occurred. Why is this evidence ambiguous?

Answer

This evidence is imprecise or ambiguous since we do not know what this observer means by the words: "tall," "very dark," and "old." These words may mean something different to her than they do to you.

36. From your own experience, can you recall other items of ambiguous evidence you have received?

Answer

How about the following: My wife told me that she is fixing something I would really like for dinner tonight. That's all she will tell me.

37. The competence, veracity, objectivity, and observational sensitivity need to be considered for all of the human sources in our dirty bomb example. Pick a human source in

this example and state what kinds of questions you would ask about the competence, veracity, objectivity, and observational sensitivity of the person you have chosen.

Answer

38. Provide an example showing how an analyst's numerical assessment of a probability applied to a conclusion can invite criticism.

Answer

Here is analyst A, whose inferential task has been to report her conclusion about which one of four competing or rival hypotheses is true: H1, H2, H3, and H4. This analysis involved a mass of evidence of different types: some HUMINT and some tangible evidence such as IMINT, SIGINT, and MASINT. Analyst A might have been assisted by other analysts who are experts in various areas relevant to this analysis. However, it is analyst A who bears responsibility for drawing a final conclusion in this important inferential assignment. Analyst A is required to provide her conclusion at a hearing involving a group of policy-makers having vital interests in the conclusion A will report.

Analyst A begins her report by reviewing the four hypotheses she has entertained in the matter of interest and the lines of evidence and arguments she has considered that bear upon these hypotheses. She concludes her report with the following comment. She says, "My judgment is that there is an 83% chance that H2 is true; as you have seen, several lines of evidence converge in favoring H2."

Hearing A's conclusion, the lead policy-maker PM1 says: "I agree that the evidence points most strongly to H2, but what troubles me is the precise probability you have used to hedge your conclusion. We know that you have not made 100 observations and discovered that the event described in H2 occurred on exactly 83 of these occasions. All the events in your hypotheses are unique and have never happened before. So, where did your precision come from? You have mentioned the array of subjective judgments you and your team of analysts had to make."

Analyst A replies: "Sir, you are correct that my conclusion was based on a variety of subjective judgments. The number 83% that I gave you is just my best estimate of this probability of H2; it could be anywhere between 75% and 87%."

Another policy-maker, PM2, then replies: "But you now give us additional precise numbers showing the upper and lower limits of your assessed probabilities. Again, where did this precision come from? This would still be based on a variety of imprecise judgments."

Analyst A then says: "I agree that these upper and lower probability estimates are imprecise; the lower limit could be anywhere between 70% and 76%, and the upper limit could be anywhere between 85% and 89%."

Hearing this response by A, the group of policy-makers are silent. Hearing no comments, A says: "The truth is that I expected you to require me to hedge my conclusion using numbers and would criticize me for hedging my conclusion in words. The second truth is that what I wished to report is that my analysis shows that H2 is quite probable and greater than the probability of any of the other hypotheses."

In response to A's final comment, the lead policy-maker PM1 says: "We actually are quite satisfied with your verbal assessment of the probability of H2. We caused you lots of trouble trying to defend the numerical assessments you made trying to justify these numbers. We agree that you have done a very fine analysis and accept your conclusion that H2 is quite probable and the most likely of any of the hypotheses you have entertained. I am going to recommend that we take a course of action consistent with H2 being true. We thank you for your very helpful analysis."

The main thing this example shows is that precise numerical probability hedges on an analytic conclusion cannot be defended because of the rampant imprecision of the judgments necessary in intelligence analysis. One virtue of probability assessments made in words is that they do not require commitments to any particular numbers when such commitments cannot be justified.

6 HYPOTHESIS TESTING

6.1 Probability Scale

We will use the symbolic probability scale from Table 4. This is a refinement of the symbolic probability scale used in the intelligence community (ICD 203), where the wide interval "55-80% likely" was split into two intervals, 55-70% likely (L) and 70-80% more than likely (ML), in order to enable more precise assessments.

Stating that hypothesis H is likely means that, based on the available evidence, H's probability of being true is between 55% and 70%. Stating that hypothesis H is more than likely means that, based on the available evidence, H's probability is between 70% and 80%. Stating that hypothesis H is lacking support means that the available evidence does not support its truthfulness with a probability greater than 50%. The probability of H could be anywhere between 0% and 50%.0

Francis Bacon

Attempting to prove some hypothesis by gathering instances favorable to it is mistaken, since all it would take to refute the generality of this hypothesis was one unfavorable instance of it.

Table 4. Probability.

Probability scale	
100%	certain (C)
95-99%	almost certain (AC)
80-95%	very likely (VL)
70-80%	more than likely (ML)
55-70%	likely (L)
50-55%	barely likely (BL)
0-50%	lacking support (LS)

6.2 Relevance of Evidence to a Hypothesis

6.2.1 Presentation and Examples

We can assess the relevance by answering the question:

Assuming that the evidence is true, what is the probability that the hypothesis is true?

When we have evidence about a fact and the hypothesis is the fact itself, the relevance of the evidence is certain, as shown in Figure 10. Indeed, if we assume that the evidence is true, then

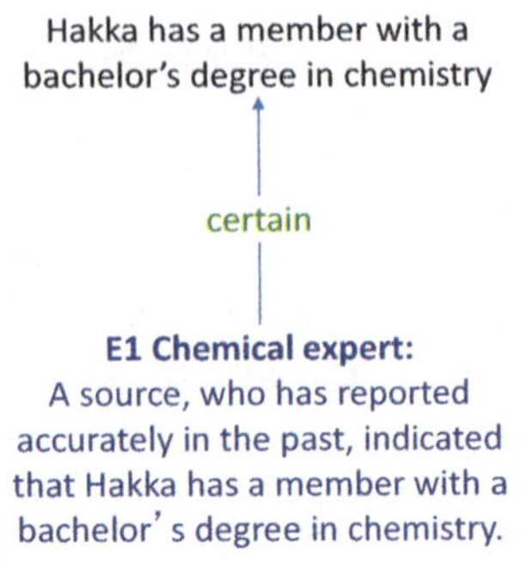

Figure 10. Simple relevance.

the hypothesis is true. But let us consider the hypothesis that "Hakka has the expertise to develop chemical weapons." If Hakka has a member with a bachelor's degree in chemistry, what is the probability that it has the expertise to develop chemical weapons?

A bachelor's program in chemistry provides basic knowledge for chemical weapons development, but this does not necessarily prove that Hakka has the expertise. Indeed, the Hakka member may not have developed this expertise. Thus, this item of evidence is not conclusive, and we assess its relevance only as likely. Explicit explanations (justifications) for assessments of relevance that are less than certain, as in this example, will clarify the reasoning and can be used in the completed argumentation to support the conclusion (see Figure 11).

When developing an argumentation, it is a good practice to consider, for each item of evidence, which is the corresponding fact, and then reason from that fact to the upper-level hypotheses, as illustrated in Figure 12.

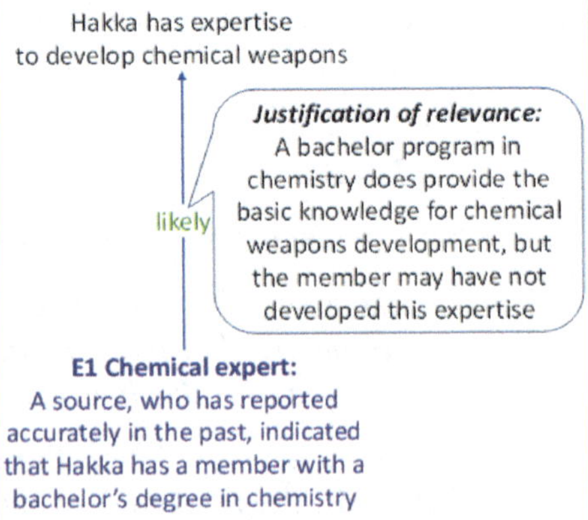

Figure 11. Justification of relevance.

The justification of the relevance assessment in Figure 11 points to the missing information that, if present, would have allowed us to assess the relevance as certain. An implied assumption has to be made for the evidence E1 to support the hypothesis. The probability of this implied assumption being true, in effect, becomes the assessed relevance of the evidence E1, as shown in Figure 12.

6.2.2 **Questions**

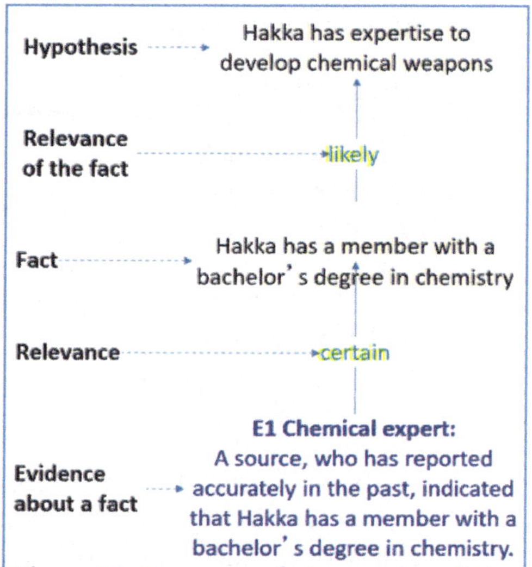

Figure 12. Recommended reasoning chain.

39. True or false? Given the hypothesis "John conducted the terrorist attack against the government on 1 May", the sub-hypothesis "John was planning to conduct a terrorist attack against the government on 1 May" should be assessed as having lower relevance than the sub-hypothesis "John was involved in several attacks against the government last year."

40. True or false?

Given the hypothesis

H: John, who is in his early 40's, is capable of embezzling money from his firm

The sub-hypothesis

H1: John was convicted of shoplifting as a teenager and should be assessed as having lower relevance than

H2: John has a reputation for having little or no integrity among most of his current co-workers.

6.3 **Credentials of Evidence**

6.3.1 **Evidence, Credibility, and Fact**

It is important to distinguish between evidence about a fact and the fact itself (see Figure 13). Consider, for example, the following item of evidence: "A source, who has reported accurately in the past, indicated that Hakka has a member with a bachelor's degree in chemistry." Can we conclude from this that Hakka has a member with a bachelor's degree in chemistry? No. The issue here is the credibility of the source, which may or may not be telling the truth.

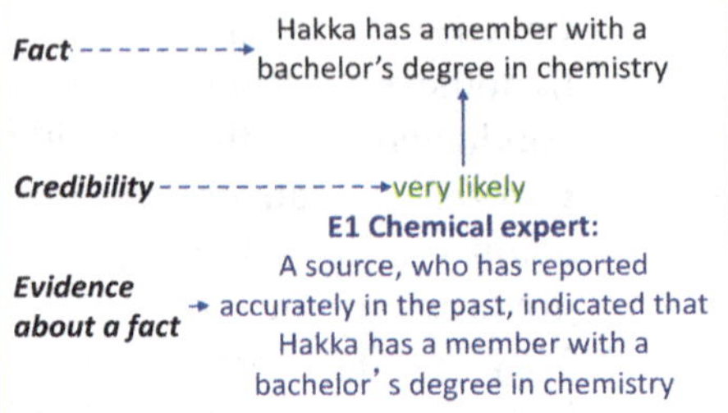

Figure 13. Evidence about a fact and the fact itself.

The credibility of an item of evidence is the extent to which the evidence may be believed. We can assess the credibility of this item of evidence by answering the following question: What is the probability that the evidence is true? This assessment can be influenced by many factors, including doubts about the source's veracity or objectivity. The source of this evidence has been reported accurately in the past. We can, therefore, assume that this current report is very likely to be true.

6.3.2 Question

41. Assuming that
    - John did not harbor any pro-terrorist views before March;
    - John visited the terrorist website 22 times in March;
    - John offered his services to terrorist group X in April.

How certain are you that

Material on the website of terrorist group X persuaded John to become a terrorist.

- certain
- almost certain

6.4 Inferential Force of Evidence

6.4.1 Presentation and Examples

The third credential of evidence is its inferential force, defined and illustrated in Figure 14. We have assessed the relevance of evidence item E1 as certain because it is an inference from evidence about a fact to the fact itself. We have also assessed the credibility of E1, which is very likely because the source has reported accurately in the past. Inferential force answers the question:

What is the probability of the hypothesis above being true based only on this item of evidence below?

In our example, the relevance of E1 is certain, but its credibility is only very likely. Therefore, the probability that Hakka does, in fact, have a member with a bachelor's degree in chemistry is only very likely.

In general, *the inferential force of an evidence item is determined as the smaller between its credibility and its relevance.* Indeed, an evidence item that is not credible would not convince us that the hypothesis is true, no matter how relevant the provided information is. Therefore, the inferential

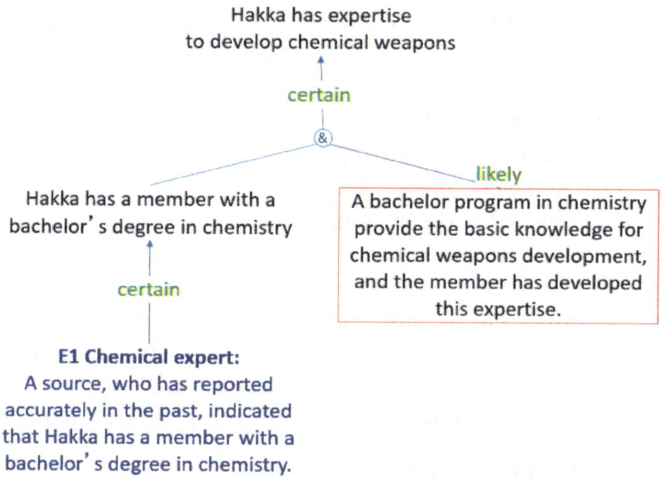

Figure 14. Relevance as pobability of an implied assumption.

force in this circumstance would be low. Similarly, it is not enough for the evidence item to be credible if the information provided is irrelevant to the hypothesis. The inferential force will be high only if the evidence item is highly relevant and credible.

In this case, because we have only one item of evidence, the probability of the hypothesis "Hakka has a member with a bachelor's degree in chemistry" is given by the inferential force of this evidence item. However, suppose we have more items of evidence, some favoring the truthfulness of the hypothesis and some disfavoring it. In that case, the probability of the hypothesis will result from the combined inferential forces of all these items of evidence.

As another example, let's now consider the upper-level hypothesis: "Hakka has the expertise to develop chemical weapons" (see Figure 15). The relevance of "Hakka has a member with a bachelor's degree in chemistry" to this hypothesis was assessed as likely. Because the probability of the sub-hypothesis is very likely, its inferential force on the top hypothesis is likely, the minimum of its relevance and probability.

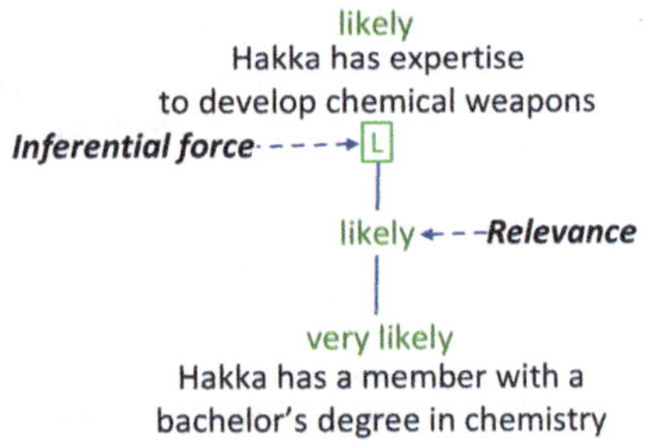

Figure 15. Another relevance assessment.

In this case, we have just this one reason and one argument for the top hypothesis to be true. Therefore, the probability of the top hypothesis is the inferential force of this argument. In general, however, we may have multiple arguments, some favoring the truthfulness of the top hypothesis and some disfavoring it. In such a case, the probability of the top hypothesis will be determined by the combined inferential forces of all these arguments.

6.4.2 **Questions**

42. A small bomb exploded in the city's center near a major university. No one was hurt, but a local monument sustained some damage. Initial police investigation indicates that the bomb had been hidden in a black backpack. Person X, a male student at the university and a member of an anti-government, anti-business anarchist

group, has been identified as a person of interest. One witness claimed to have seen Person X near the bombing site just before the explosion, while another witness claimed that Person X was elsewhere at the time.

43. Consider the following hypothesis and evidence:

H: Person X was in the vicinity at the time of the bombing.

E_1^* : Eyewitness 1 claims to have seen Person X near the bombing site shortly before the bombing occurred. This eyewitness had never met or seen Person X before.

E_2^*: Eyewitness 2 claims that Person X was at another location at the time of the bombing. Eyewitness 2 is a member of the same anarchist group.

44. Which of the following statements is correct?
    E_1^* favors the truthfulness of H.
    E_1^* disfavors the truthfulness of H.
    E_1^* is not relevant to H.

45. Which of the following values is a more accurate assessment of the relevance of E_1^*:
    certain (100%)
    very likely (80-95%)
    barely likely (50-55%)

46. Which of the following values is a more accurate assessment of the credibility of E_1^*:
    almost certain (95-99%)
    likely (55-70%)

47. Which of the following statements is correct?
    E_2^* favors the truthfulness of H.
    E_2^* disfavors the truthfulness of H.
    E_2^* is not relevant to H.

48. Which of the following values is a more accurate assessment of the relevance of E_2^*:
    certain (100%)
    very likely (80-95%)
    barely likely (50-55%)

49. Which of the following values is a more accurate assessment of the credibility of E_2^*:
    almost certain (95-99%)
    barely likely (50-55%)

6.5 Evidence-Based Hypothesis Assessment

6.5.1 Presentation and Examples

In general, there may be several items of evidence that are relevant to a given hypothesis, some favoring it and some disfavoring it, each with a specific credibility, relevance, and inferential force. Figure 16 illustrates a situation where there are two items of evidence favoring hypothesis H (E1 and E2), and one item of evidence disfavoring it (E3). Each item of evidence has a specific inferential force on hypothesis H given by the smallest between its credibility and relevance. Thus, the inferential force of E1 is likely, that of E2 is almost certain, and that of E3 is likely.

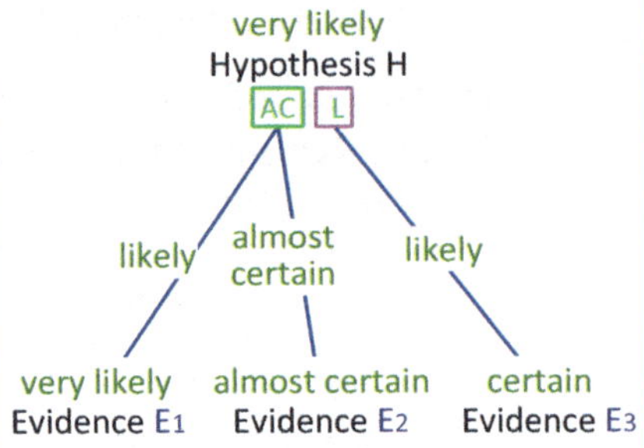

Figure 16. Hypothesis assessment based on the credentials of evidence.

The combined inferential force of the two favoring items of evidence (E1 and E2) is almost certain, the highest between the inferential force of E1 and that of E2. This combined inferential force is displayed in the left (green) box under hypothesis H to indicate that it favors the truthfulness of H. The inferential force of the disfavoring item of evidence (E3) is likely. This combined inferential force is displayed in the right (pink) box under hypothesis H to indicate that it disfavors the truthfulness of H.

The probability of hypothesis H is determined by balancing the combined inferential force of the favoring evidence (almost certain) and the inferential force of the disfavoring evidence (likely).

The Baconian probability view (Cohen, 1977; 1989) requires considering either H or not H as probably true, but not both simultaneously. To assess a hypothesis that has both favoring and disfavoring evidence, such as hypothesis H in Figure 16, we have introduced an on-balance function shown in Table 5 that balances the inferential force of the favoring evidence (almost certain) with that of the disfavoring evidence (likely), assessing a probability of more than likely for H.

Table 5. On-balance function.

Inferential force of disfavoring evidence/arguments (columns); Inferential force of favoring evidence/arguments (rows)

H	lacking support	barely likely	likely	more than likely	very likely	almost certain	certain
lacking support	lacking support	lacking support	lacking support	lacking support	lacking support	lacking support	lacking support
barely likely	barely likely	lacking support	lacking support	lacking support	lacking support	lacking support	lacking support
likely	likely	barely likely	lacking support	lacking support	lacking support	lacking support	lacking support
more than likely	more than likely	likely	barely likely	lacking support	lacking support	lacking support	lacking support
very likely	very likely	more than likely	likely	barely likely	lacking support	lacking support	lacking support
almost certain	almost certain	very likely	more than likely	likely	barely likely	lacking support	lacking support
certain	certain	almost certain	very likely	more than likely	likely	barely likely	lacking support

As indicated in the right and upper sides of Table 5, if the inferential force of the disfavoring evidence is higher than or equal to that of the favoring evidence, then hypothesis H lacks support. If, however, the inferential force of the favoring evidence is strictly greater than that of the disfavoring evidence (and there is some force of the disfavoring evidence), then the probability of H is lowered, based on the inferential force of the disfavoring evidence (see the left and lower side of Table 5).

6.5.2 **Questions**

50. Which of the following is an assumption in the argument

that John has a higher IQ than Mark:

- John scored higher than Mark on the SAT.
- Individuals who score higher on the SAT have a higher IQ.
- John scored higher than Mark on an IQ test.

51. Possible answers to a question about a situation are considered:
    - assumptions
    - hypotheses
    - items of evidence

52. Consider the hypothesis "Mark's grades have improved" and the sub-hypotheses:
    - Last year, Mark maintained a C average.
    - Mark is maintaining a B average this year.

How should you represent them?

a) As two alternative (separate) arguments
b) As a single AND argument

53. Consider the hypothesis "John, who is in his early 40s, is capable of embezzling money from his firm" and the sub-hypotheses "John was convicted of shoplifting as a teenager" and "John has a reputation of having little or no integrity among most of his current co-workers."

How should you represent them?

a) As two alternative (separate) arguments
b) As a single AND argument

6.6 Assessing Complex Hypotheses

The previous sections presented a simple way of directly assessing a hypothesis based on evidence. This works well when it is easy to assess the relevance of the evidence to the hypothesis. However, it does not work well for assessing complex hypotheses. Such complex hypotheses are decomposed into simpler and simpler hypotheses, down to the level of very simple hypotheses for which the relevance of evidence can be more confidently assessed, as discussed in this section.

6.6.1 Simple Argument

Let us consider the hypothesis "Hakka has chemical weapons." What condition would make this hypothesis true? IF Hakka develops chemical weapons, THEN Hakka has chemical weapons.

You also need to assess the relevance of the sub-hypothesis to the hypothesis by answering the question: Assuming that Hakka does indeed develop chemical weapons, what is the probability that Hakka has chemical weapons?

If Hakka develops chemical weapons, then we would expect it to have chemical weapons. However, there is a small chance that it may not be successful in producing functional chemical weapons. Therefore, you may assess the relevance of the "develops" sub-hypothesis as "almost certain." When assessing the relevance, it is recommended that you record the justification of your assessment in the argument (see Figure 17).

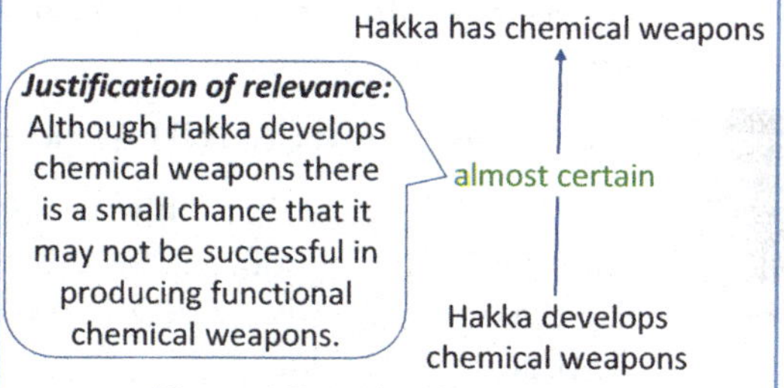

Figure 17. A simple argument with justification of relevance.

Once the probability of the sub-hypothesis is determined (for example, likely), the system computes its inferential force as the smallest between its probability and its relevance. Thus, the inferential force of this simple argument is likely. Because this is the only argument for the top hypothesis, the probability of the top hypothesis is given by the inferential force of this argument.

6.6.2 Alternative (OR) Arguments

Is there any other condition that would make the top hypothesis true? Yes, IF Hakka buys chemical weapons THEN it is certain that it has chemical weapons. This is an alternative argument for the top hypothesis "Hakka has chemical weapons" (see Figure

18).

Let us assume that the probability of "Hakka buys chemical weapons" is barely likely. Then, the inferential force of this alternative argument is also barely likely, the smallest between the probability of the sub-hypothesis and its relevance. Now we have two arguments for the truthfulness of "Hakka has chemical weapons," one with inferential force likely, and the other with inferential force barely likely. In such a case, the system assesses the probability of the top hypothesis being the highest of the two inferential forces. Thus, it is likely that "Hakka has chemical weapons" (see Figure 18). Here is a general rule: *The probability is determined by the highest inferential force of the individual supporting arguments.*

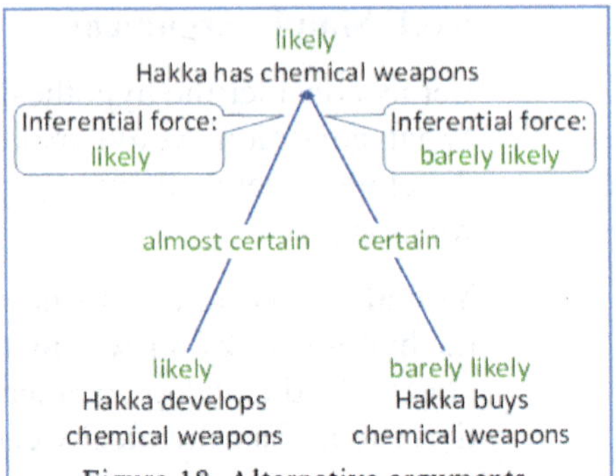

Figure 18. Alternative arguments.

6.6.3 **Conjunctive (AND) Argument**

It may be the case that multiple sub-hypotheses need to be true in order for the hypothesis to be true, as in the example from: IF Hakka has expertise AND production material AND funds, THEN Hakka develops chemical weapons.

These three sub-hypotheses, taken together, represent an argument for the top hypothesis. You also need to assess the relevance of this AND argument, and it is

Figure 19. An AND argument.

recommended that you also justify your assessment of this relevance: There may be several additional conditions that are necessary to develop chemical weapons. Therefore, based on these indicators alone, we assess that it is only very likely that Hakka develops chemical weapons. The probability of the top hypothesis is determined by the inferential force of the AND argument, which is the smallest among the probabilities of the sub-hypotheses and the relevance of the argument, as shown in Figure 19.

6.6.4 Disfavoring Evidence and Arguments

If you are rigorously developing and evaluating your argumentation, you will be taking into account both favoring and disfavoring evidence and arguments for a given hypothesis. Let us consider the hypothesis: "Hakka buys chemical weapons."

A favoring argument or reason for this hypothesis (see Figure 20), whose relevance we judge as very likely, is: "Hakka has funds." However, there would be strong sanctions against the seller, and we judge that this almost certainly will deter any potential seller.

Consider that it is very likely Hakka has funds. Then, the inferential force of this favoring argument is computed as very likely. Now consider that, almost certain, there will be strong

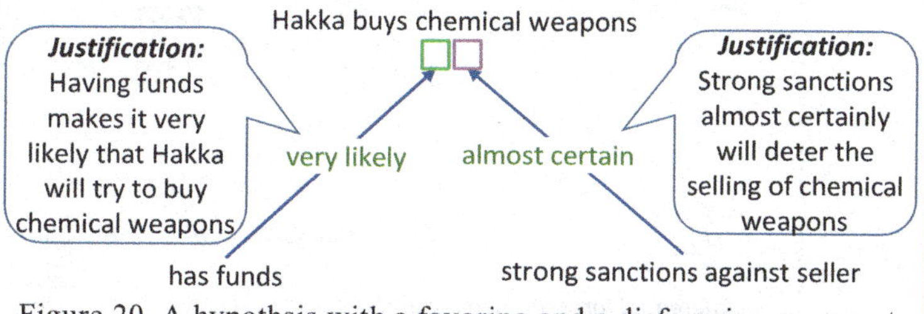

Figure 20. A hypothsis with a favoring and a disfavoring argument.

sanctions against the seller. The inferential force of this disfavoring argument is computed as almost certain. As a result, based on the on-balance function from Table 5, the hypothesis "Hakka buys chemical weapons" is lacking support. Therefore, the evidence does not support the conclusion that Hakka buys chemical weapons, but it supports the conclusion that it does not buy them.

6.6.5 Summary of Wigmorean Argumentations

Figure 21 summarizes the hypothesis assessment process using an abstract example of an augmented Wigmorean argumentation, which is a probabilistic inferential network integrating logic with Baconian and Fuzzy probabilities. Hypothesis H is decomposed into simpler hypotheses by considering both favoring arguments (supporting the truthfulness of H), under the left (green) square and disfavoring arguments (supporting the falsehood of H under the right (pink) square.

Each argument is an independent strategy to show H is true or false and is characterized by a specific relevance or strength. The argument consists either of a single hypothesis (e.g., H_3) or a conjunction of hypotheses (e.g., H_1 & H_2). The sub-

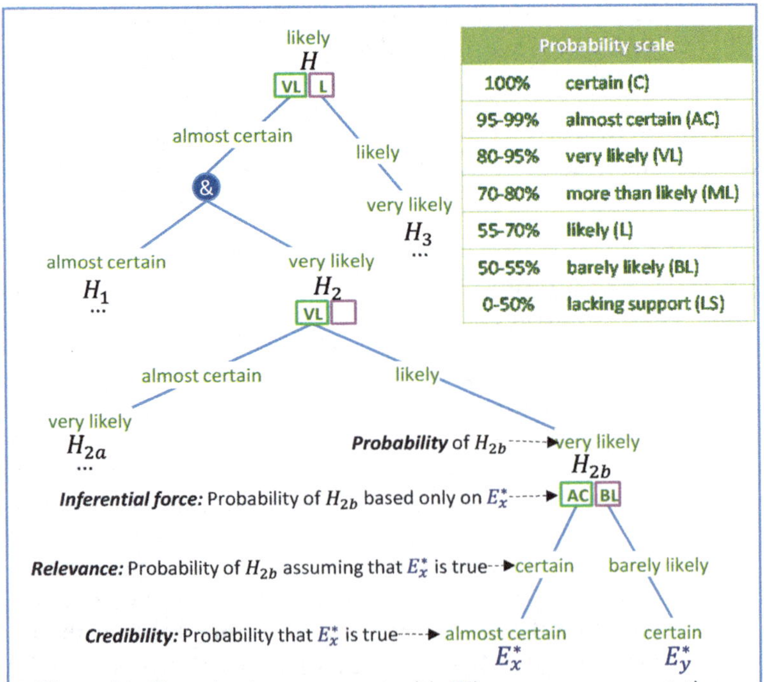

Probability scale	
100%	certain (C)
95-99%	almost certain (AC)
80-95%	very likely (VL)
70-80%	more than likely (ML)
55-70%	likely (L)
50-55%	barely likely (BL)
0-50%	lacking support (LS)

Figure 21. Hypothesis assessment with Wigmorean argumentation.

hypotheses from these arguments are further decomposed through other arguments, leading to simpler and simpler (sub-sub-)hypotheses that can be more accurately assessed based on evidence.

Consider, for example, sub-sub-hypothesis H_{2b}. There are two items of evidence relevant to this hypothesis, the favoring evidence item E_x^*, and the disfavoring evidence item E_y^*.Each item of evidence has three credentials that need to be assessed: credibility, relevance, and inferential force.

The credibility of evidence answers the question: *What is the probability that the evidence is true?*

The relevance of evidence to a hypothesis answers the question: *What would the probability of the hypothesis be if the evidence were true?*

Based on these two credentials, the system computes the inferential force or weight of the evidence on the hypothesis that answers the question: *What is the probability of the hypothesis, based only on this evidence?* This is computed as the minimum between credibility and relevance. For example, the inferential force of E_x^* is almost certain (100%), that of E_y^* is barely likely (50-55%).

The probability of sub-sub-hypothesis H_{2b} is determined as very likely (80-95%) by balancing the inferential force of the favoring evidence with that of the disfavoring evidence. Therefore the probability of H is likely, based on the on balance function from Table 5.

Once the probabilities of the bottom-level hypotheses have been determined based on evidence, the probabilities of the upper-level hypotheses are computed as explained below, based on the logical structure of the Wigmorean argumentation (conjunctions and disjunctions of hypotheses), using min-max probability combination rules common to the Fuzzy probability view (Zadeh, 1983; Negoita and Ralescu, 1975; Schum 2001) and the Baconian probability view (Cohen, 1977; 1989; Schum, 2001). These rules are much simpler than the Bayes rule used in the Bayesian probability view (Schum, 2001) or the Dempster-

Shafer rule in the Belief Functions probability view (Shafer, 1976).

Hypothesis H_2 has two favoring arguments, H_{2a} with inferential force very likely (80-95%) and H_{2b} with inferential force very likely (80-95%). Its probability is very likely (80-95%), the maximum between the inferential force of H_{2a} (very likely, 80-95%) and that of H_{2b} (likely, 55-70%). Finally, hypothesis H has a favoring argument with inferential force very likely (the minimum of almost certain, almost certain, and very likely) and a disfavoring argument with inferential force likely (the minimum of very likely and likely). Balancing these probabilities results in the probability of H as being assessed as likely.

6.7 Answers

39. True or false?

Given the hypothesis "John conducted the terrorist attack against the government on 1 May", the sub-hypothesis "John was planning to conduct a terrorist attack against the government on 1 May" should be assessed as having lower relevance than the sub-hypothesis "John was involved in several attacks against the government last year."

Answer

False: The relevance of "John was planning to conduct a terrorist attack against the government on 1 May" should be higher because that sub-hypothesis very specifically connects John to the 1 May attack. In contrast, the information that "John was involved in several attacks against the government last year" does not state anything conclusively about what John is doing or planning regarding the 1 May attack.

40. True or false?

Given the hypothesis

H: John, who is in his early 40's, is capable of embezzling money from his firm

The sub-hypothesis

H1: John was convicted of shoplifting as a teenager

should be assessed as having lower relevance than

H2: John has a reputation for having little or no integrity among most of his current co-workers.

Answer

True. The sub-hypothesis "John was convicted of shoplifting as a teenager" should be assessed as having lower relevance. The relevance of this sub-hypothesis speaks to John's moral compass as a youth, not his current moral compass. It is possible that John has turned his life around. In contrast, the sub-hypothesis that "John has a reputation of having little or no integrity among most of his current co-workers" speaks to his current moral compass and suggests few of his co-workers trust John to do the right thing.

41. Assuming that
    - John did not harbor any pro-terrorist views prior to March.
    - John visited the terrorist website 22 times in March.
    - John offered his services to terrorist group X in April.

How certain are you that "Material on the website of terrorist group X persuaded John to become a terrorist"?

a) certain
b) almost certain

Answer

Almost certain is a better answer than certain because we do not know exactly what John thought about the material on the website. It appears the material on the website was highly influential, but we cannot be 100% certain. It is possible that he harbored some doubts about the material on the website but was influenced to act by other factors.

44. A small bomb exploded in the city's center near a major university. No one was hurt, but a local monument sustained some damage. Initial police investigation indicates that the bomb had been hidden in a black backpack. Person X, a male student at the university and a member of an anti-government, anti-business anarchist group, has been identified as a person of interest. One

witness claimed to have seen Person X near the bombing site just before the explosion, while another witness claimed that Person X was elsewhere at the time.

Consider the following hypothesis and evidence:

H: Person X was in the vicinity at the time of the bombing.

E_1^*: Eyewitness 1 claims to have seen Person X near the bombing site shortly before the bombing occurred. This eyewitness had never met or seen Person X before.

E_2^*: Eyewitness 2 claims that Person X was at another location at the time of the bombing. Eyewitness 2 is a member of the same anarchist group.

Which of the following statements is correct?

E_1^* favors the truthfulness of H.
E_1^* disfavors the truthfulness of H.
E_1^* is not relevant to H.

Answer

If Person X was near the bombing site shortly before the bombing occurred, then it is expected that he was still in the vicinity at the time of the bombing. Therefore, E_1^* favors the truthfulness of H.

43. Which of the following values is a more accurate assessment of the relevance of E_1^*:

certain (100%)
very likely (80-95%)
barely likely (50-55%)

Answer

If Person X was near the bombing site shortly before the bombing occurred, then very likely is a more accurate estimate than certain or barely likely that he was still in the vicinity at the time of the bombing.

44. Which of the following values is a more accurate assessment of the credibility of E_1^*:

almost certain (95-99%)

likely (55-70%)

Answer

Eyewitness 1 does not know Person X, and eyewitnesses are often mistaken. Therefore, the credibility is less than almost certain, and a more accurate assessment is likely.

45. Which of the following statements is correct?

E_2^* favors the truthfulness of H.
E_2^* disfavors the truthfulness of H.
E_2^* is not relevant to H.

Answer

If Person X was at another location at the time of the bombing, then he was not in the vicinity at the time of the bombing. Therefore, E_2^* disfavors the truthfulness of H.

46. Which of the following values is a more accurate assessment of the relevance of E_2^*:
certain (100%)
very likely (80-95%)
barely likely (50-55%)

Answer

If Person X was near the bombing site shortly before the bombing occurred, then very likely is a more accurate estimate than certain or barely likely that he was still in the vicinity at the time of the bombing.

47. Which of the following values is a more accurate assessment of the credibility of E_2^*:
almost certain (95-99%)
likely (55-70%)

Answer

Eyewitness 1 does not know Person X, and eyewitnesses are often mistaken. Therefore, the credibility is less than almost certain, and a more accurate assessment is likely.

7 PRACTICUM IN ARGUMENT CONSTRUCTION

7.1 Introduction

Benjamin Franklin
Tell me and I forget, teach me and I may remember, involve me and I learn.

It appears that analysts are so often trained in the production of intelligence analyses rather than in the actual process of analysis itself. Very little training is offered regarding the properties, uses, discovery, and marshaling of the evidence upon which all analyses rest. As Douglas MacEachin (1999), former DDI, CIA, remarked, *"Information and expertise are a necessary but not sufficient means of making intelligence analysis the special product that it needs to be. A comparable effort has to be devoted to the science of analysis."*

Analysis of evidence cannot be learned effectively just by listening to someone discuss his/her own analyses or just by giving students lectures and assigned readings on the topics. What is absolutely necessary is a regular practice involving analyses of evidence using either hypothetical situations or examples drawn from actual situations. In short, evidential analysis is mastered best by performing analyses contrived to illustrate the wide variety of subtleties or complexities so often encountered in actual evidential analyses.

Analysts need all the help they can get in the face of a tsunami of information and the requirement to answer questions of immediate interest that do not allow time for extensive research on and deliberation of available evidence. Therefore, analysis tools based on solid theoretical foundations should be integrated into intelligence analysis courses to teach the process of analysis (including the properties, uses, discovery, and marshaling of the evidence) through a learning-by-doing approach. Moreover, given the above characteristics of intelligence analysis, we believe that it can be best performed through the mixed-initiative integration of human imagination and computer knowledge-based reasoning.

7.2 Cognitive Assistant for Cogent Analysis

Cogent is an intelligent system that assists an intelligence analyst in solving intelligence analysis problems (http://lac.gmu.edu/Cogent/index.html). It builds on a series of analytical tools that include Disciple-LTA (Tecuci et al., 2008), TIACRITIS (Tecuci et al., 2011), and Disciple-CD (Tecuci et al., 2016a). To install Cogent on your computer, follow the instructions corresponding to your operating system.

7.2.1 Cogent Setup on macOS Computers

Cogent-v2022.08.1m has the following recommended configuration:

- MacBook Pro or iMac 2016 or later
- 8GB RAM
- 150MB available on hard disk (more space is required if Java is not installed)
- 15" or larger monitor with a screen resolution of at least 1920x1200
- macOS Big Sur
- Java 17 or later (https://www.oracle.com/java /technologies/downloads/#jdk17-mac)
- After downloading the most recent Java distribution file from the link above, double-click it and follow the displayed installation instructions.
- Download Cogent from http://lac.gmu.edu/cogent/Cogent-v2022.08.1m.zip.
- Unzip the distribution file (Cogent-v2022.08.1m.zip) in your home folder (/Users//Cogent-v2022.08.1m.zip; for example /Users/JohnSmith/Cogent-v2022.08.1m).
- If you chose to unzip the distribution file in a different location, you must edit the second line in the 'Cogent.command' file to point to the actual location relative to your home folder (for example, if you unzipped the distribution file on the desktop, you must change the second line to 'cd Desktop/Cogent-

v2022.08.1m').

- You are now ready to use Cogent by double-clicking the file Cogent.command in the distribution folder (Cogent-v2022.08.1m).
- If you obtain a security error and are unable to open the program with a double click, then perform the following operations:
- Open a Finder window in your home folder
- Navigate to the folder 'Cogent-v2022.08.1m' and select it
- CONTROL-CLICK on the Cogent.command file
- Select the Open command
- Override the security message and select to Open the program

7.2.2 **Cogent Setup on Windows Computers**

Cogent-v2022.08.1w has the following recommended configuration:

- Intel Core processor (2.4GHz)
- 8GB RAM
- 150MB available on hard disk (more space is required if Java is not installed)
- 15" or larger monitor with a screen resolution of 1920x1200
- Microsoft Windows 10
- Java 17 from https://download.oracle.com/java/17/latest/jdk-17_windows-x64_bin.msi
  After downloading the most recent Java distribution file from the link above, double-click it and follow the displayed installation instructions.
- Download Cogent for Windows from http://lac.gmu.edu/cogent/Cogent-v2022.08.1w.zip.
- Unzip the distribution file (Cogent-v2022.08.1w.zip) in the desired location (for example, in your home folder 'C:\Users\').
- You are now ready to use Cogent by double-clicking the file Cogent.bat in the distribution folder (Cogent-v2022.08.1w).

7.2.3 Cogent Data

Cogent stores analyses in your home folder inside the folder structure 'CogentData/Cogent/repository'. When you run Cogent for the first time, it creates this folder structure in your home folder. If you download and use future versions of Cogent, please DO NOT delete "CogentData" because the already existing analyses from this folder will be used by the new versions of Cogent.

7.2.4 Uninstall Cogent

To uninstall Cogent-v2022.08.1 Delete the distribution folder ('Cogent-v2022.08.1w' on Windows or 'Cogent-v2022.08.1m' on macOS)

If you plan to install a new version of Cogent, skip this step; otherwise, if you plan to no longer use Cogent, delete the data folder from your home folder (CogentData).

When you delete the data folder, all the analyses developed with Cogent will be lost. Make a copy of the 'CogentData\Cogent\repository' folder if you need to preserve this data.

7.2.5 Documentation

Cogent has a knowledge base named "Getting Started". It is a step-by-step guide for basic operations in Cogent. You can load this knowledge base by clicking on "Menu" -> "Open Analysis" and then double-clicking on "Getting Started". On the bottom left of the screen, there is an up arrow. Clicking on that can expand the instructions. Additionally, if you click on a particular step of the instructions, it provides details and examples of this step.

7.3 Assessing the Relevance and Credibility

The relevance of an item of evidence to a hypothesis indicates how strongly does this item support a specific hypothesis. We assess the relevance by answering the question:

> *Assuming that the evidence is true, what is the probability that the hypothesis is true?*

The relevance of an item of evidence to a hypothesis is certain when it is essentially a complete and exact reflection of what is in the hypothesis, as illustrated in Figure 22.

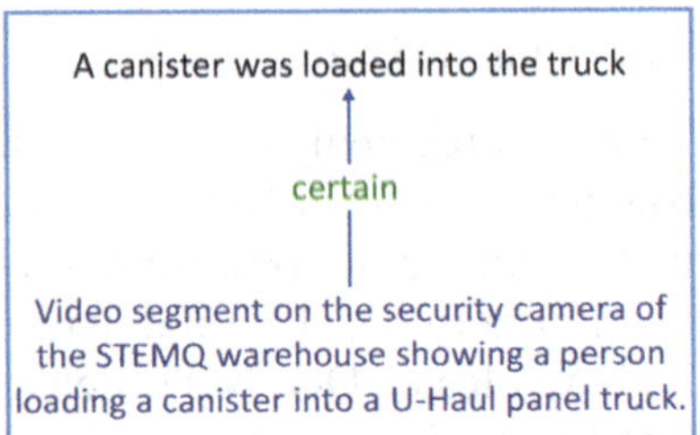

Figure 22. Relevance of evidence.

The relevance of incomplete evidence is

Table 6. Credibility asessment guidane.

Evidence	Credibility assessment
Encrypted intercepted comm.	At least almost certain (95-99%). This information is encrypted because people are trying to hide it from others.
Intercepted communications	Almost certain (95-99%) if there is no contradictory information, the individual has no reason to lie, and he/she is in a position to know. Otherwise: barely likely (50-55%) or even lacking support (0-50%).
Imagery	Certain (100%) unless the imagery is incomplete or not clear, perhaps because of cloud cover. The credibility of imagery, for example, of 70 percent of an air base that shows 10 fighters present is certain that there are at least 10 and likely (55-70%) or more than likely (70-80%) that there are only 10.
Technical measurements from radars, space sensors, etc.	Certain (100%) or almost certain (95-99%) if the measurement equipment is operating properly. The credibility is adjusted downward depending on the extent to which the measurement equipment is not operating properly. For example, the credibility of information from a radar that is only 60 percent accurately calibrated is only likely (55-70%).
Reporting from human sources	Very likely (80-95%) if the source has access, has a track record of reporting accurate information, and has no reason to be biased or lying. Otherwise, the credibility has to be discounted downward. Can reach as low as barely likely (50-55%) or lacking support (0-50%).
Documentary information: technical documents, court records, police records, forensic evidence, etc.	Certain (100%) or almost certain (95-99%) if the source of the information, if it is a statement, has no reason to lie, and the source in a position to know. Information from witnesses to a crime that are not close to the defendant or victim, for example, would be certain (100%) or almost certain (95-99%). Technical documentation or forensic evidence that is not subject to misinterpretation would be certain.
Media and press reporting	Certain (100%) if there is direct coverage of the statement, such as video coverage of a speech or a statement made in an interview. If there is any suspicion of bias in coverage or doubt about whether the media is in a position to know, credibility has to be discounted downward, perhaps as low as lacking support (0-50%).

given by its degree of completeness, as illustrated in Figure 23.

The relevance of an argument to a hypothesis is certain when the argument is essentially a complete and exact reflection of what is in the hypothesis, as illustrated in Figure 24. The relevance of an argument to a hypothesis is less than certain when the argument is ambiguous or inconclusive.

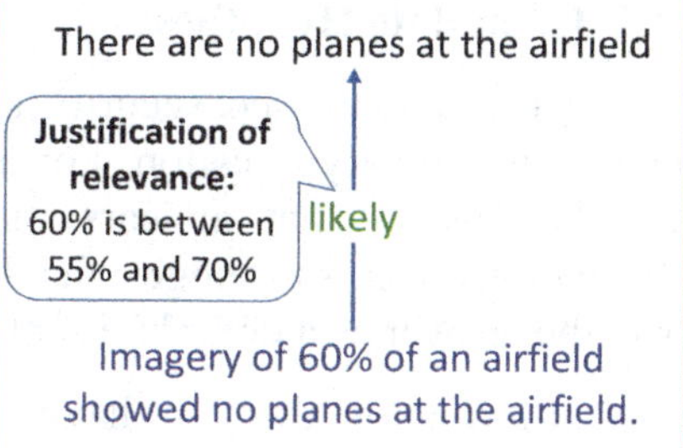

Figure 23. Relevance of an incomplete item of evidence

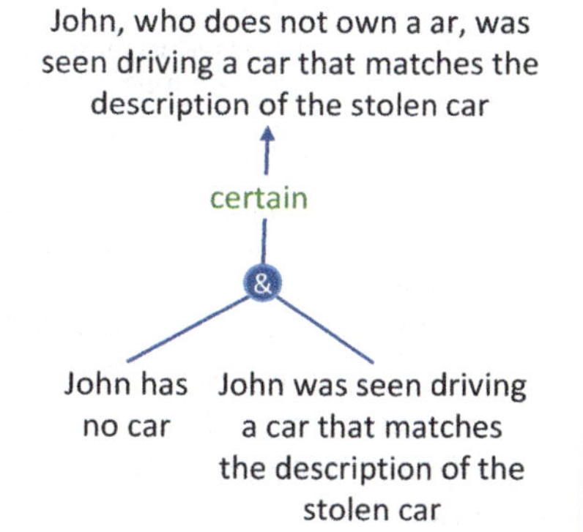

Figure 24. Relevance of argument.

7.4 Power Plant Security

Consider the problem in Table 7. The following illustrates a general strategy to answer the intelligence question.

Table 7. Power plant security problem.

Situation: The United States has passed information to the country Upland that terrorists are planning to attack Upland's nuclear power plants in order to seize radioactive material for a "dirty" bomb. The United States is concerned that Upland is not taking this information seriously and that some plants are still vulnerable to attack.
Question: Has Upland increased security at all of its nuclear power plants?
Available Information: In an intercepted message, the Power Minister told the Prime Minister that additional fences and security cameras were installed at all nuclear power plants in June.
A newspaper report said that security has been upgraded at all nuclear power plants. Imagery shows that new fences were installed at 3 nuclear power plants. Imagery of the other plant was not available.

7.4.1 Formulate Hypotheses

"Has Upland increased security at all of its nuclear power plants?" is a binary question. For a binary question, consider first the answer that appears more likely. The available information appears to suggest that Upland has, but you should write the answer as a complete sentence rather than "Yes."

What information do you have that suggests Upland has increased security at all of its nuclear power plants?

- In an intercepted message, the Power Minister told the Prime Minister that additional fences and security cameras were installed at all nuclear power plants in June.
- A newspaper report said that security had been upgraded at all nuclear power plants.
- Imagery shows that new fences were installed at three nuclear power plants. Imagery of the other plant was not available.

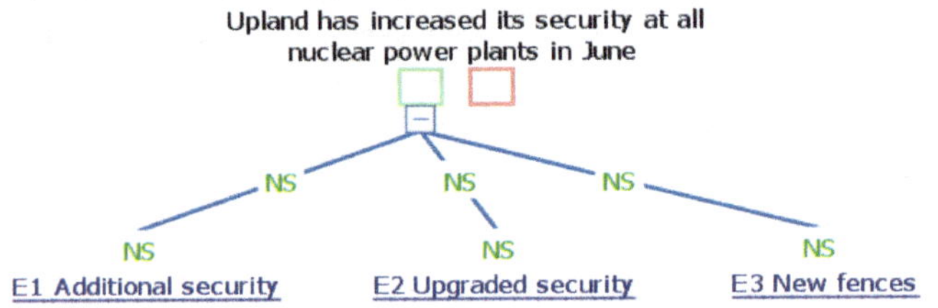

Figure 25. Initial arguments.

Any one of the above pieces of information can support the hypothesis independently of the others. They will become the basis for three alternative arguments, as shown in Figure 25.

7.4.2 Evaluate Relevance

The relevance of E1 is certain because if E1 is true, the hypothesis certainly is true.

The relevance of E2 is certain because if E2 is true, the hypothesis certainly is true.

The relevance of E3 is more than likely (70-80%) because that information relates to only three of Upland's four nuclear power plants, which is 0.75.

7.4.3 Evaluate Credibility

The credibility of E1 is almost certain. This information is from an intercepted communication and we have no reason to believe the Power Minister was lying.

The credibility of E2 is barely likely; the information is from a newspaper, and we have no information on how or where the newspaper acquired this information.

The credibility of E3 is certain because the information is from imagery and provides visual proof of improved security at three power plants.

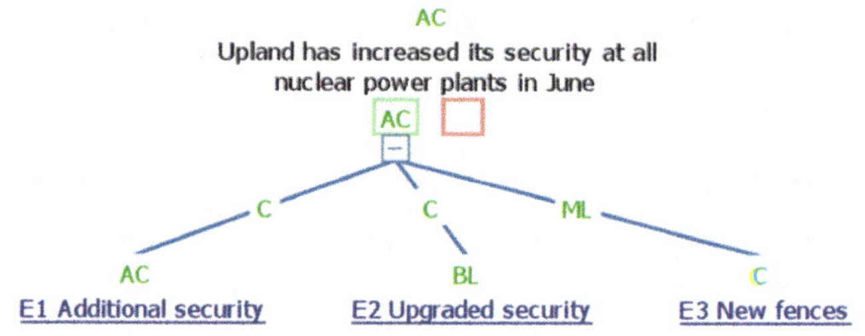

Figure 26. Final arguments.

7.5 Economics Minister Fired

Consider the problem in Table 8.

Table 8. The eonomics minister fired problem.

Situation: The Prime Minister has fired the Economics Minister.

Question: Why did the Prime Minister fire the Economics Minister?

Available Information: A reliable source who is a longtime aide to the Prime Minister said that the Prime Minister values complete loyalty above all else. The Economics Minister told the President that the Prime Minister was incompetent and lazy, according to a reliable source on the Prime Minister's staff.

7.5.1 Formulate Hypotheses

Do you have any information that suggests why did the Prime Minister fire the Economics Minister? Yes:

- The Prime Minister values complete loyalty above all

else.

- The Economics Minister told the President that the Prime Minister was incompetent and lazy.

This information will become the basis for the argument in Figue 27. This is an "and" argument because both of the sub-hypotheses below must be true for the hypothesis above to be true.

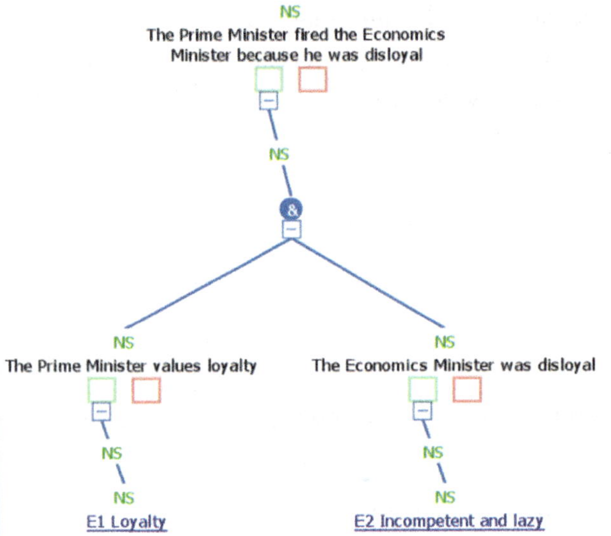

Figure 27. Initial arguments.

7.5.2 **Evaluate Relevance**

The relevance of E1 is assessed as certain because if the evidence is true, the hypothesis above it must be true.

The relevance of E2 is assessed as almost certain because we are not certain how the Prime Minister defines "loyalty." It is possible that the Prime Minister accepts some personal criticism as long as his instructions on policy are completely carried out.

7.5.3 **Evaluate Credibility**

We assess the credibility of both E1 and E2 as very likely because both of the human sources are assessed as "reliable," and both have good access to the information in question.

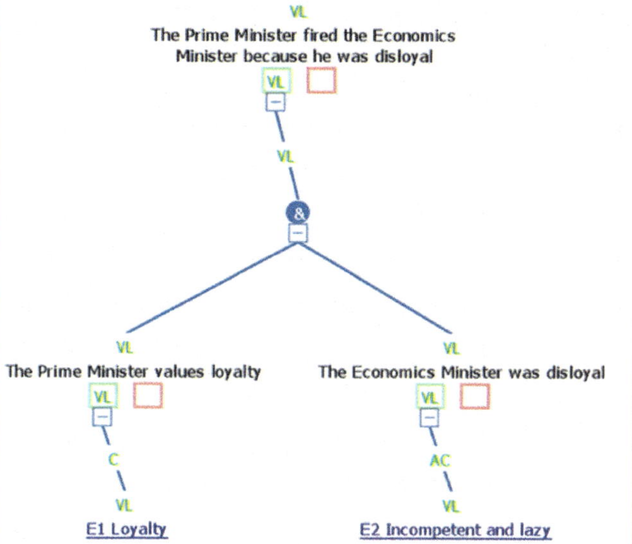

Figure 28. Final arguments.

7.6 Law School

Consider now the problem in Table 9

Table 9. The law school problem.

Situation: John is graduating from college and his parents are wondering what John will do next.
Question: Will John apply to law school?
Available Information: John told his girlfriend he has the legal mind of a Supreme Court justice. John, according to numerous comments in his year book, is the only student at his college that wears a tie and suit to class. John told his brother that he wants to be making at least $150,000 a year by the time he is 30.

7.6.1 Identify Hypotheses

Look for information that is relevant and might allow you to answer the question.

Do you have any information that suggests John will apply to law school? Yes:

- John has the legal mind of a Supreme Court justice.

- John is the only student at his college who wears a tie and suit to class.
- John wants to be making at least \$150,000 a year by the time he is 30.

This information will become the basis for the argument I Figure 29.

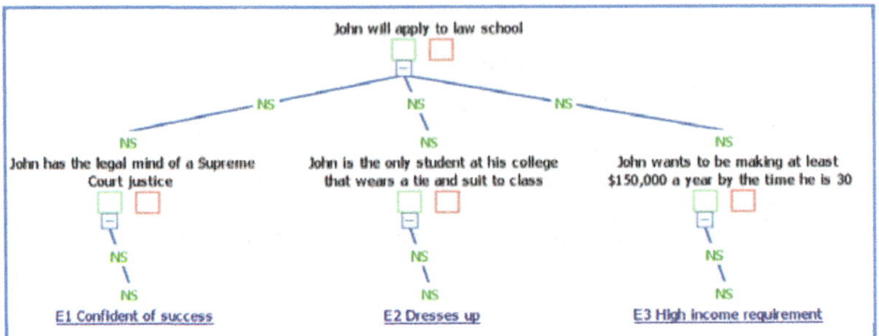

Figure 29. Initial arguments.

7.6.2 **Evaluate Relevance**

The relevance of the first sub-hypothesis is assessed as very likely. Most people choose a profession that they like and think they would be good at. However, there may be other professions that John may believe he would be good at.

The relevance of each of the next two arguments is assessed as barely likely because there are many other professions that pay well and require you to wear a coat and tie.

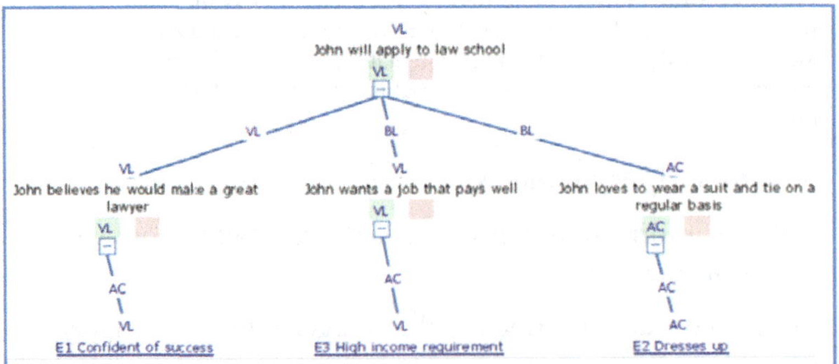

Figure 30.Final arguments.

7.6.3 Evaluate Credibility

The credibility of E1 is assessed as very likely because most people who are dating will have talked about their next steps as they approach graduation and possible separation.

The credibility of E2 is assessed as almost certain because numerous people separately made this observation.

The credibility of E3 is assessed as very likely because John's brother is in a position to know.

7.7 Entrance Exam

Finally, consider the problem in Table 10.

Table 10. The entrance exam problem.

Situation: Nick wants to go to law school. He will have to do well on his law school entrance exams because he barely graduated as an undergraduate with a 2.02 GPA.

Question: How will Nick do on the law school entrance exam?

Available Information: His IQ was measured at 145. Nick has been working at the car wash 60 hours a week for the past three months, according to his employer's records. He scored a 1475 on the SATS and 1490 on the GRE. Nick's dad said that Nick, who now lives at home, is partying all the time when he isn't working.

7.7.1 Identify Hypotheses

Will Nick do well on the law school entrance exam? Yes, his IQ was measured at 145.

Do you have any information that suggests Nick will not do well on the law school entrance exam? Yes.

Nick has been working at the car wash 60 hours a week for the past three months, according to his employer's records.

Nick's dad said that Nick, who now lives at home, is partying all the time when he isn't working.

This information will become the basis for the argument in Figure 31.

7.7.2 Evaluate Relevance and Credibility

The relevance of E2 is almost certain: The tests as a measure of a person's IQ are reliable but not perfect. If his IQ, which is a measure of intelligence, is high, we can conclude for certain that Nick is intelligent.

The relevance of this favoring argument is likely. Nick is intelligent, but there is no guarantee that he will do well on the test. Some questions may be based on specific knowledge. The credibility of E2 is almost certain because these standardized tests seldom confuse one person's results with another's.

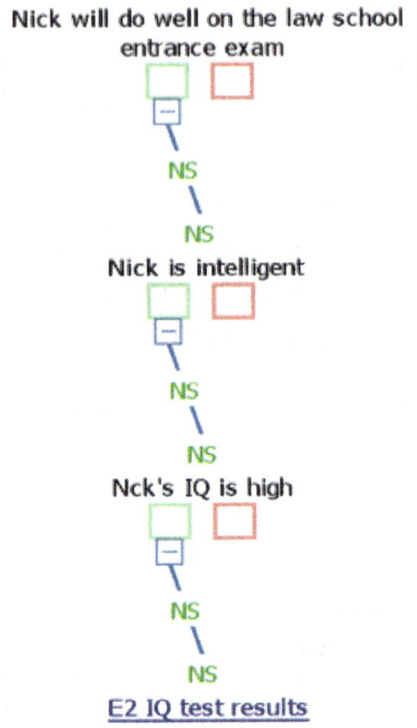

Figure 31. Initial arguments.

The relevance of E3 is certain because we can conclude for certain that Nick is working 60 hours a week. The relevance of E4 is certain because we can conclude for certain that Nick is partying when he is not working. This is an "and" argument because the sub-hypothesis above is a statement whether Nick has time to prepare for the test.

The relevance of the argument below is assessed as very likely: Nick is spending his time either working or partying, but we don't know much about his job; maybe he has time to study while he works. The relevance of this disfavoring argument is barely likely. Some specific knowledge is needed but the test is largely a "thinking" exam.

The credibility of E3 is assessed as likely. Nick's dad is in a position to know much about what Nick is doing but not everything. The credibility of E4 is certain because it is based on documentary employee records.

7.7.3 Insert Disfavoring Arguments and Evidence

We now have an argument and different sub-hypotheses for which there may be counter sub-hypotheses and counter-evidence.

Do you have any information that suggests Nick is not intelligent? Yes, Nik barely graduated as an undergraduate with a 2.02 GPA

Do you have any information that suggests Nick is preparing enough for the exam? Yes, He scored a 1475 on the SATS and 1490 on the GRE.

7.7.4 Evaluate Relevance

The relevance of E1 to the sub-hypothesis "did poorly in school" is also certain. "Barely graduated" is arguably synonymous with "did poorly." The relevance of this disfavoring argument is assessed as barely likely because many intelligent people do poorly in school.

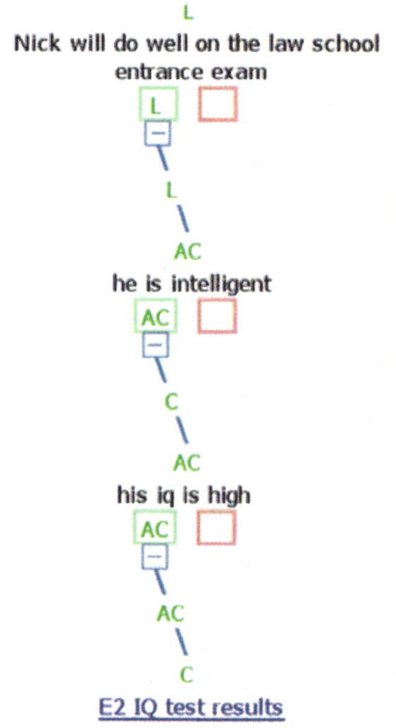

Figure 32. Final arguments.

The relevance of E5 is assessed as likely: maybe Nick got lucky on the SAT and GRE. We assessed the relevance to the hypothesis that Nick is not preparing enough as likely. The law school entrance exam is a different test, and doing well on some standardized tests is not guaranteed that he will do well on other tests.

7.7.5 Evaluate Credibility

The credibility of his undergraduate record (E1) is assessed as certain because this is indisputable documentary information. That of E5 is certain because it is documentary evidence.

Figure 33. Final arguments.

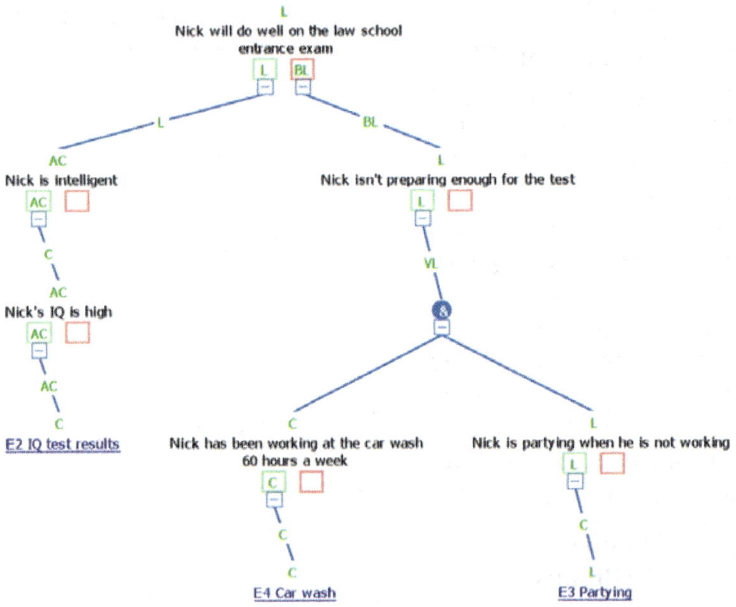

Figure 34.Final arguments.

8 PRACTICAL APPLICATIONS

Jacques Hadamard
Practical application isfound by not looking for it, and one can say that the whole progressof civilization rests on that principle.

In the previous chapters, we presented how to perform critical thinking in everyday activities. However, the same process can be followed to solve complex, real-world problems, as illustrated in this chapter.

8.1 Intelligence, Surveillance and Reconnaisance

8.1.1 Intrduction

Intelligence, Surveillance, and Reconnaissance (ISR) synchronizes and integrates the planning and the operation of collection assets with processing, exploitation, and dissemination systems in direct support of current and future operations (DOD JP 1-02, 2010). Sensemaking is the process of situational understanding based on data that is sparse, noisy, and uncertain (Moore, 2011).

With support from the U.S. Department of Defense, we have developed a proof of concept system called Multi-Agent System for Automatic Sensemaking through Hypothesis Generation and Analysis (MASH). MASH can be instructed to perform a wide variety of ISR tasks, such as:

- Automatically monitor plants and facilities on a global basis to determine construction or operating status, including production facilities as well as transportation infrastructure, such as pipelines, ports, and loading terminals.
- Automatically assess the intended purpose of facilities under construction worldwide as new information comes in and construction advances.
- Automatically assess indicators for determining the likelihood of whether an attack or explosion occurred at plants and facilities on global basis.
- Automatically monitor changes in security at industrial

and military facilities on a global basis.

- Automatically assess the country's capability to support the production of various weapon systems.
- Automatically assess indicators of threat perceptions for a multitude of countries.

8.1.2 **Hypothesis Generation**

Through abductive (imaginative) reasoning, which shows that something is possibly true, the analyst and MASH generate hypotheses that may explain an observation. The problem is that, in general, there are very many such hypotheses, as shown

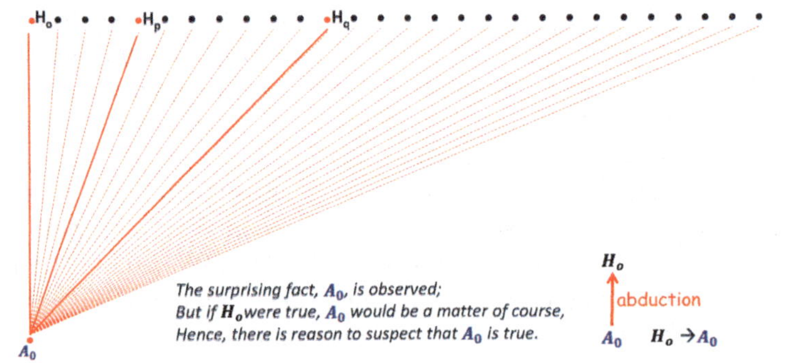

Figure 35. Hypohesis generation through abductive reasoning.

by the red dots at the top of Figure 35.

We developed a much more efficient approach where a hypothesis is generated through a chain of abductive steps $A_o \rightarrow F_i \rightarrow G_l \rightarrow H_0$. Each step involves the generation and testing of competing hypotheses to determine the most likely one that will be used in the next abductive step (see Figure 36):

$$\text{If } A_o \text{ then } possibly\ F_i, possibly\ F_j, \text{ and } possibly\ F_k$$

After that, evidence is discovered for F_i , F_j and F_k, and the probabilities of these hypotheses are assessed. The process continues with the next abduction step from the most likely hypothesis (e.g., F_i):

$$\text{If } F_i \text{ then } possibly\ G_l,\ possibly\ G_m, \text{ and } possibly\ G_n$$

Then, each of these hypotheses is assessed based on the

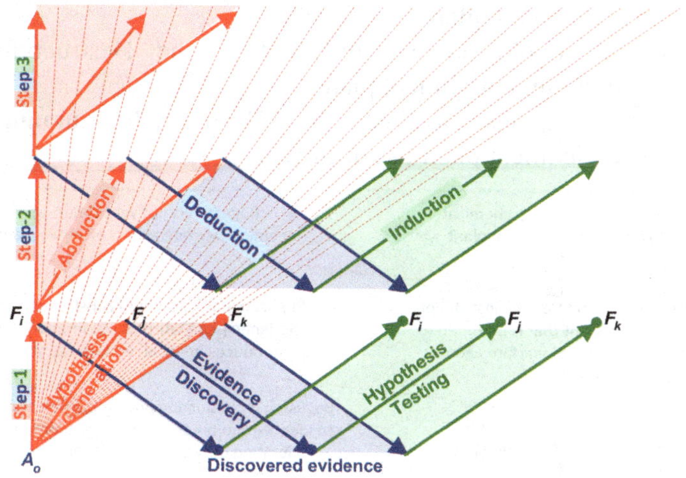

Figure 36. Hypothesis generataion though multi-sep abdution, deducion, and induction.

discovered evidence. If G_i is the most likely one, then the process continues with the next abduction step. From G_l, we would abduce the hypotheses of interest H_o, H_p, and H_q, and assess them.

This approach based on spiral hybrid reasoning, where small abductive, deductive, and inductive steps feed each other, significantly speeds-up hypothesis generation.

To illustrate, consider the scenario in Table 11.

Table 11. The Bogustan scenario.

Bogustan was building a new chemical plant at Tanan that was nearing completion; the plant's purpose was not known. Bogustan was suspected of harboring weapons of mass destruction ambitions.
A reconnaissance asset conducting a routine quarterly overflight detected heat signatures at the Tanan facility.

The automatic ISR process starts with an alerting observation:

> According to a thermal imagery sensor using heat detection, several areas of the Destructville uranium enrichment plant are emitting heat as of 3/5/2020.

The first step of the multi-step abduction process shown in Figure 37 results in the following competing hypotheses:

- Equipment is operating at the Destructville uranium enrichment plant.
- An explosion has occurred at the Destructville uranium enrichment plant.
- An attack against the Destructville uranium enrichment plant has occurred.

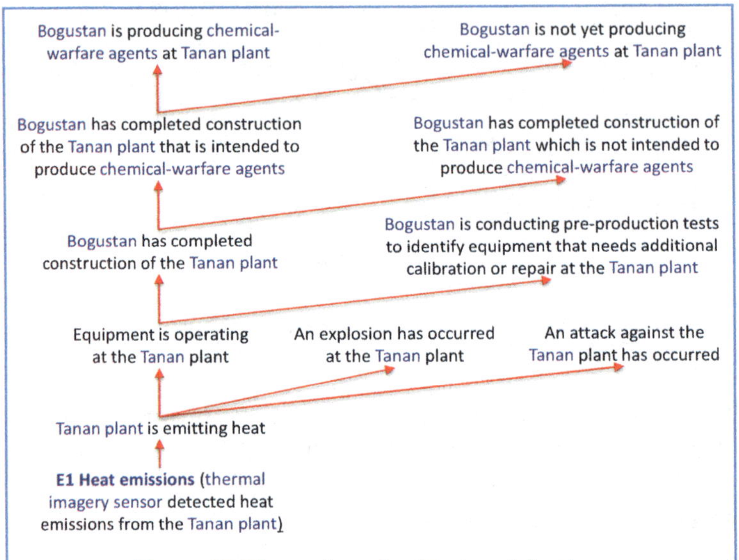

Figure 37.Illustration ofmulti-step abduction.

8.1.3 Evidence Discovery

To determine which of these competing hypotheses is true, the system uses each hypothesis and deductive reasoning, which shows that something is necessarily true, to discover new evidence. The question is: *What evidence would favor or disfavor this hypothesis?*

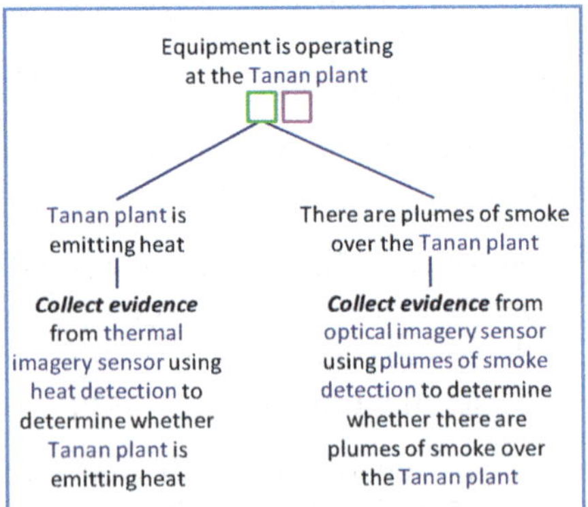

Figure 38.Evidence discovery.

MASH first identifies sufficient conditions or

indicators for or against each hypothesis and then looks for evidence related to these conditions and indicators that it would expect to observe (see Figure 38).

8.1.4 **Hypothesis Testing**

Once the evidence is discovered, the agent employs inductive reasoning (that shows that something is probably true) to determine the probability of the hypothesis. This process is repeated for each step of the abduction-deduction-induction process, as illustrated in Figure 39).

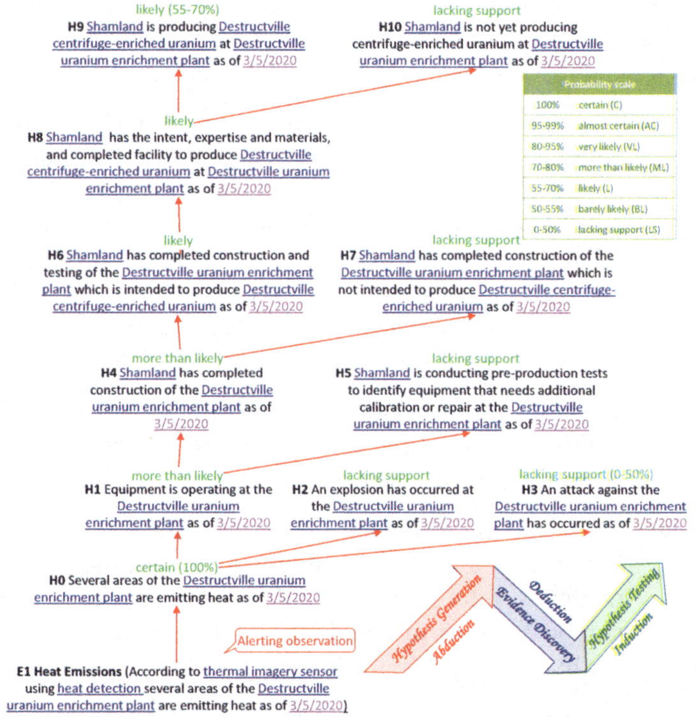

Figure 39. Multi step abducion, deduction, and induction.

8.1.5 **Comparison with the State of the Art**

The prevailing approach to critical thinking in intelligence analysis and in many other domains is a holistic one where the analyst, after reviewing large amounts of information and mentally processing the data, reaches a conclusion (Marrin, 2011). Consequently, there is very limited transparency on how exactly the conclusion has been reached from evidence, what assumptions have been made, and how exactly the probability of the conclusion and the confidence in this probability has been assessed.

Complementary to the holistic approach is the structured analysis approach, but the current practice relies on very simple analytic techniques, such as those described in (Heuer and Pherson, 2015).

However, in his laudable attempt to simplify the complex analysis task, Heuer made several over-simplifications that affected the result of the analysis process, such as:

- The credibility of evidence is assessed holistically, without considering that different types of evidence have different credentials. For example, the credibility of a drone image depends on its accuracy and the reliability of the camera, while the credibility of an intelligence report depends on the source's competence, veracity, objectivity, and observational sensitivity, as discussed in (Tecuci and Schum, 2023).
- The relevance of evidence is also assessed holistically without constructing relevant arguments necessary for more accurate assessments.
- The selection of the most likely hypothesis is ad hoc. For example, a variant of the ACH method assigns numeric values to credibility and relevance and computes sums of "credibility * relevance" to select the hypothesis with the highest value.

There is no assessment of the probability of a hypothesis. The highly acclaimed book on Critical Thinking by Pherson and Pherson (2021) presents general guidelines for good analysis but no method for assessing the probability of a hypothesis.

Much more advanced methods use Bayesian probabilistic inference networks, but despite their implementation in advanced analytical tools, such as Netica developing, a

Bayesian network is a difficult task for an intelligence analyst (see https://www.norsys.com/).

Great advances in sensor technology have made it very easy to automatically collect massive amount of data, leading to the so-called "deluge of data." This huge and growing gap between the analysts' ability to collect information and that of analyzing it, is making increasingly difficult for them to draw timely and accurate conclusions from these massive amounts of collected data, much of which is irrelevant to the hypotheses of interest.

Among the most advanced analytical tools is Cogent, a cognitive agent for intelligence analysis (Tecuci et al., 2015; 2018), which is the latest in a series of analytical tools that include Disciple-LTA (Tecuci et al., 2007), TIACRITIS (Tecuci et al., 2011), and Disciple-CD (Tecuci et al., 2016a).

8.1.6 **Further Reading**

Tecuci, G., Marcu, D., Boicu, M, Kaiser, L. (2020). Instructing a Cognitive Agent to Perform Sensemaking in Intelligence, Surveillance and Reconnaissance, Eighth Annual Conference on Advances in Cognitive Systems (ACS 2020), online, Palo Alto Research Center, Palo Alto, CA, August 10-12.

Tecuci, G., Marcu, D., Kaiser, L., Boicu, M. (2021). Shared Model of Sense-making for Human-Machine Collaboration, The 2021 AAAI Fall Symposium "AI in Government and Public Sector," Arlington, VA, Nov.4-6.

8.2 Cybersecurity

8.2.1 Introduction

Despite the tens of billions of dollars spent on cybersecurity every year, the number of successful attacks has increased steadily year over year due to the fast evolution of attacker methodology.

Cyber defense is done in a Cybersecurity Operations Center (CSOC) where analysts monitor alerts and log data from available information sources, each having differing levels of credibility, and use them to determine the presence or absence of intrusion activity. However, because a single alert alone does not provide sufficient evidence that an incident has occurred, and modern detection technologies are error-prone, each alert must be carefully examined and validated by a human analyst. Both missed detections, and false positives are costly, resulting in real damage to an organization or wasted incident response resources, respectively. In a large enterprise, thousands of alerts can be reported daily. Therefore, even sensors with a false positive rate of one percent may have enough missed detections and false positives to be unmanageable by even mature CSOCs.

Among the most sophisticated cyber threats faced by an organization are those known as Advanced Persistent Threats (APTs). These are computer network exploitation groups (many of them state-sponsored) that leverage superior resources, knowledge, and tactics to gain and maintain access to targeted networks and adapt to defenders' efforts to resist them.

APT1 is the name given to a group of APT actors, attributed to China's People's Liberation Army unit 61398, who led a years-long campaign of cyber espionage dating back to at least 2004. APT1 is known for a regimented approach to computer intrusion activity. An APT1 intrusion typically consists of the following phases:

- Gain access to a network by sending fraudulent, malicious email messages to specific users (spearphishing);

- Use multiple types of backdoor programs to maintain presence and provide remote connectivity to the target network;
- Use a collection of command-and-control (C2) servers to obfuscate the source of their attacks;
- Escalate privileges and acquire legitimate login credentials to access network resources;
- Move laterally within the target network using legitimate credentials to gain redundant points of presence and identify information of interest and
- Exfiltrate targeted information through their C2 infrastructure.

With support from the U.S. Air Force Research Laboratory, we have developed a proof-of-concept system for APT detection.

8.2.2 Hypotheses Generation

APT detection requires reasoning over a large set of weak indicators of compromise, such as unique strings, filenames, and Registry keys, that must be aggregated to infer the presence of an intrusion. For example, consider the alert from the bottom of Figure 40, a suspicious connection to our computer from an external computer using an APT1 domain known to belong to the Chinese military. The three hypotheses that may explain this alert are at the top of the figure.

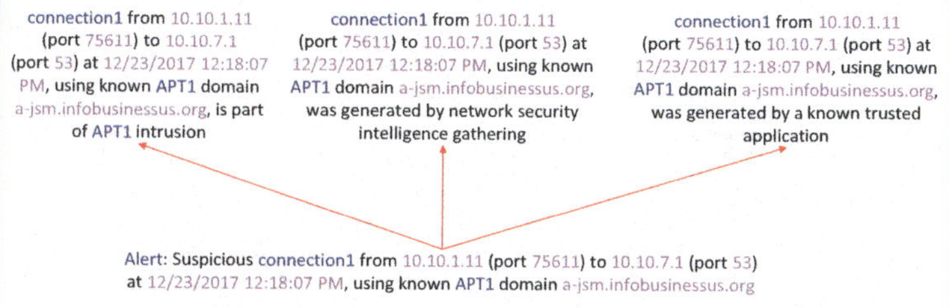

Figure 40. Hypothesis generation.

8.2.3 Evidence Discovery

Each generated hypothesis is used to guide the collection of additional evidence, as illustrated in Figure 41.

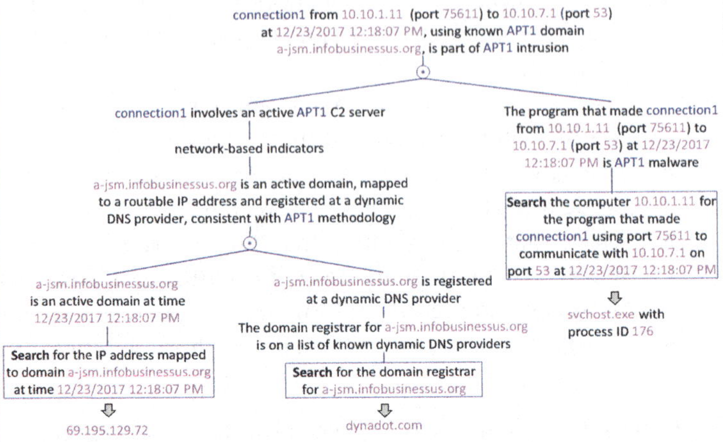

Figure 41. Evience discovery.

8.2.4 Hypothesis Testing

This illustrates the hypothesis analysis process. There are two leading indicators of the top hypothesis:

- connection1 involves an APT1 command and control server (APT1 C2 server),
- The program that made connection1 is an APT1 malware (APT1 malware).

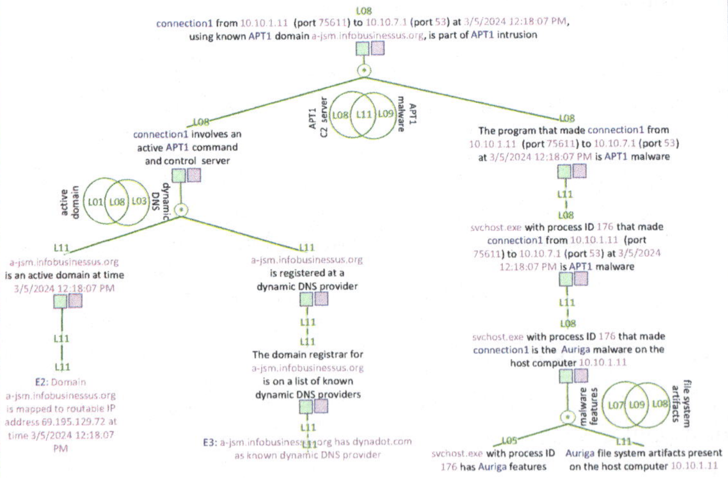

Figure 42. Hypothesis testing.

The problem is that, in a given situation, the agent may find evidence for the presence of only the first indicator, the second one, or the presence of both. The * operator corresponds to the disjunction of these three possibilities. The probabilities inside the two overlapping circles show these three cases' relevance to the top hypothesis's truthfulness. Notice also that in this domain, we used a refined probability scale.

Table 12. Extended Probabilities

L11	100%	certain
L10	95-99%	almost certain
L09	90-95%	very likely+
L08	85-90%	very likely
L07	80-85%	very likely-
L06	75-80%	more than likely+
L05	70-75%	more than likely
L04	65-70%	likely+
L03	60-65%	likely
L02	55-60%	likely-
L01	50-55%	barely likely
L00	0-50%	lacking support

8.2.5 **Further Reading**

Meckl, S., Tecuci, G., Boicu, M., Marcu, D. (2015). Towards an Operational Semantic Theory of Cyber Defense Against Advanced Persistent Threats, Proceedings of the 10th International Conference on Semantic Technologies for Intelligence, Defense, and Security - STIDS, Fairfax, VA, November 18-20.

Meckl, S., Tecuci, G., Marcu, D., Boicu, M., Zaman, A.B. (2017). Collaborative Cognitive Assistants for Advanced Persistent Threat Detection, Proceedings of the 2017 AAAI Fall Symposium "Cognitive Assistance in Government and Public Sector Applications," pp.171-178, Arlington, VA, November 9-11, Technical Reports FS-17-01-FS-17-05, AAAI Press: Palo Alto, CA.

Tecuci, G., Meckl, S., Marcu, D., Boicu, M. (2018). Evidence-based Detection of Advanced Persistent Threats, Special Issue on Evidence-based Reasoning and Applications, *Computing in Science and Engineering*, 20(6), pp.54-65, Nov./Dec.

Tecuci, G., Meckl, S., Marcu, D., Boicu, M. (2019). Instructable Cognitive Agents for Autonomous Evidence-Based Reasoning, Advances in Cognitive Systems, 8. Also in Proceedings of the Seventh Annual Conference on Advances in Cognitive Systems, Technical Report COLAB2-

TR-4, pp.183-204, Massachusetts Institute of Technology, Cambridge, MA, August 2-5.

Huang, J., An, Z., Meckl, S., Tecuci, G., Marcu, D. (2020). Complementary Approaches to Instructable Agents for Advanced Persistent Threats Detection, Studies in Informatics and Control, 29(3). Impact Factor 2019: 2.102.

Meckl, S., Tecuci, G., Marcu, D., Boicu, M. (2021). Automating the Investigation of Sophisticated Cyber Threats with Cognitive Agents, in Dasgupta, P., Collins, J., Mittu, R. (Eds.) Adversary-Aware Learning Techniques and Trends in Cybersecurity, DOI 10.1007/978-3-030-55692-1, ISBN 978-3-030-55691-4, Springer International Publishing, pp.125-146.

8.3 Antipatory Intelligence

Anticipatory intelligence is the complex task of identifying and assessing new, emerging trends, changing conditions, and underappreciated developments to challenge long-standing assumptions, encourage new perspectives, identify new opportunities, and provide warning of threats to national interests based on current information of all kinds that come from a variety of different sources (ODNI, 2019). It is aimed at potential events, including low-probability and high-threat events, and involves active attention management, focusing on likely critical information sources (Klein et al., 2007).

8.3.1 An Example of Anticipatory Analysis

The following example shows how evidence about a missing cesium-137 canister leads to anticipating that a dirty bomb will be set off in the Washington, D.C., area. Note that this scenario and all the entities involved are fictitious.

Mavis reads an article in today's Washington Gazette concerning how safely radioactive materials are stored in this area. Willard, the investigative reporter and author of this piece, begins by noting how the storage of nuclear and radioactive materials is so frequently haphazard in other countries and wonders how carefully these materials are guarded here in the USA, particularly in this general area. In the process of his investigations, the reporter notes his discovery that a canister containing cesium-137 had gone missing from the STEMQ Company in Maryland just three days ago. The STEMQ Company manufactures devices for sterilizing medical equipment and uses cesium-137 in these devices along with other radioactive materials.

8.3.2 Hypothesis Generation

The missing cesium-137 canister arouses Mavis's curiosity because she is concerned about terrorists planting dirty bombs in our cities. The question is:

What hypotheses would explain this observation?

She experiences a flash of insight that a dirty bomb may be set off in the Baltimore-Washington, D.C. area (see **Error!**

Reference source not found.). However, no matter how imaginative or important this hypothesis is, no one will take it seriously unless Mavis can justify it. She needs to build an argument linking the evidence to her hypothesis. In doing this, she is assisted by the six honest serving men from "The Elephant's Child" poem by Rudyard Kipling [1865 - 1936]:

I kept six honest serving men,
(They taught me all I knew),
Their names are What and Why and When
and How and Where and Who.

So, she develops the chain of abductive inferences shown on the left side of Figure 43:

- There is evidence that the cesium-137 canister is missing, so it is possible that it is indeed missing.
- It is further possible that it was stolen, possibly by a terrorist organization.
- The terrorist organization could use the cesium-137 canister to build a dirty bomb and set it off in the Washington, D.C., area.

However, these are not the only hypotheses that may explain the evidence. Just because there is evidence that the cesium-137 canister is missing does not mean that it is indeed missing. The issue here is the credibility of the source of this information. Thus, an alternative hypothesis is that the cesium-137 canister is not missing.

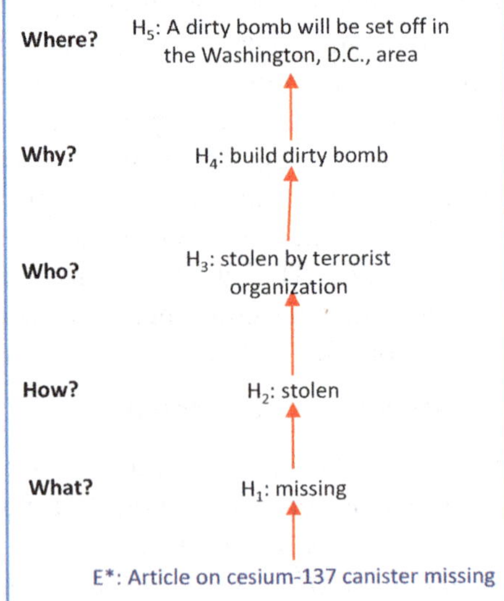

Figure 43. Chain of abductive inferences.

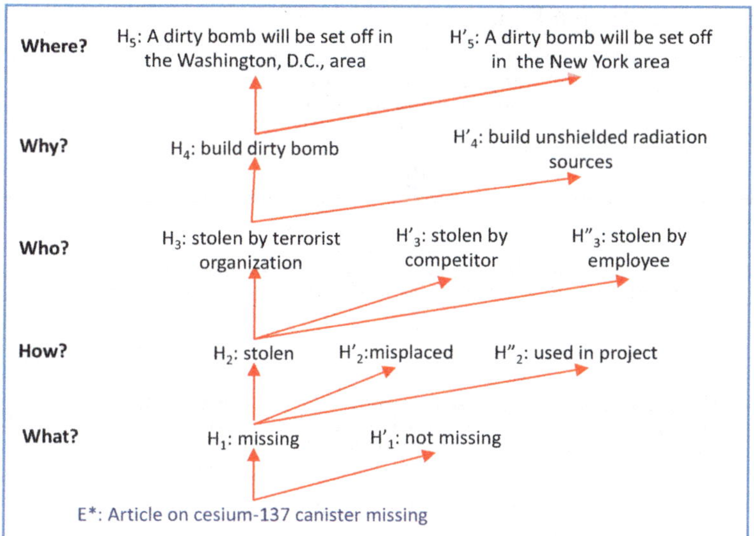

Figure 44. Competing hypotheses.

But let us assume that it is missing. Then it is possible that it was stolen, but it is also possible that it was misplaced, or maybe it was used in a project at the STEMQ Company. Now, let us assume that the cesium-137 canister was stolen. Then, it is possible that it was stolen by a terrorist organization, by a competitor of STEMQ, or by an employee. Upper-level hypotheses concern possible future events, such as the terrorist organization will use the cesium-137 canister to build a dirty bomb, the dirty bomb will be set off in the Washington, D.C., area, or it will be set off in the New York area.

8.3.3 Evidence Discovery

Mavis needs to assess each of these competing hypotheses and determine which of them is more likely. Starting from the bottom up, each hypothesis is put to work to guide the collection of additional evidence:

- Assuming the cesium-137 canister is missing, what other things should be observable?
- What are the necessary conditions for an object to be missing from a warehouse? It was in the warehouse; it is no longer there, and no one has checked it out, as

shown in Figure 45.

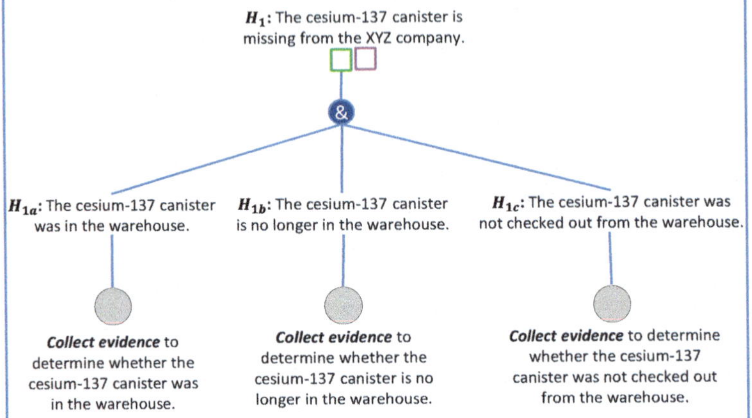

Figure 45. Evidence discovery.

As a result, the analyst contacts Ralph, the warehouse supervisor, who reports that the cesium-137 canister is registered as being in the warehouse, that no one at the STEMQ Company had checked it out, and that it is not located anywhere in the hazardous materials locker. He also indicates that the lock on the hazardous materials locker appears to have been forced.

8.3.4 **Hypothesis Testing**

Ralph's testimony provides several items of relevant evidence, and the question is: What is the probability that the cesium-137 canister is missing, based on this evidence? To answer this question, Mavis builds the Wigmorean probabilistic inference network from Figure 46.

First, based on the corresponding relevant evidence, she must assess the probabilities of the bottom hypotheses in Figure 54. Then, these probabilities are composed to produce the probability of the top hypothesis.

The credibility of the evidence item E2 Canister registered is

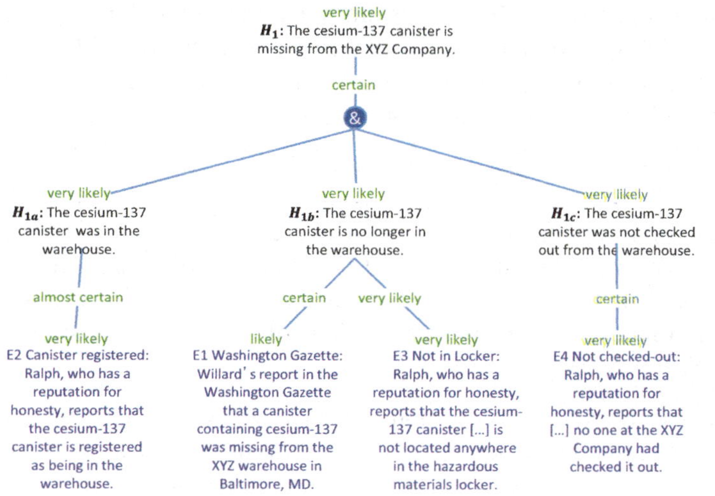

Figure 46. Wigmorean probabilistic inference network.

assessed as very likely because its source, Ralph, has access to the reported information and has a reputation for honesty. The relevance of E2 Canister registered is assessed as almost certain because the records of the STEMQ Company are almost certainly correct. Consistent with both the Baconian and the Fuzzy min/max probability combination rules, the inferential force of E2 Canister registered on the hypothesis H1a is determined as the minimum between the credibility of E2 (very likely) and the relevance of E2 (almost certain). Thus, the inferential force of E2 on H1a is very likely.

As shown at the top of Figure 46, it is very likely that the cesium-137 canister is missing, this being the minimum between the probabilities of the three sub-hypotheses and the relevance of their conjunctive argument.

Some of the newly discovered evidence may trigger new hypotheses or the refinement of the current hypotheses. For example, during her initial investigation, Mavis discovered a video segment from a security camera at the warehouse showing a person loading a container into a U-Haul truck, leading her to refine the "stolen" hypothesis to indicate that the cesium-137 canister was stolen with a U-Haul truck.

Having concluded that the cesium-137 canister is missing, Mavis and the cognitive assistant must now establish whether it was stolen with a truck, misplaced, or used in a project at the STEMQ Company. Each of these hypotheses is used to guide the collection of evidence for assessing it: If the cesium-137 canister was stolen with a truck, what other things should be observable?

Based on the current evidence, Mavis imagines the following scenario on how the cesium-137 canister might have been stolen: The truck entered the company, the canister was stolen from the locker, the canister was loaded into the truck, and the truck left with the canister.

Such scenarios have enormous heuristic value in advancing the investigation because they consist of mixtures of what is taken to be factual and what is conjectural. Conjecture is necessary to fill in natural gaps left by the absence of existing evidence. Each such conjecture, however, opens up new avenues of investigation and the discovery of additional evidence if the scenario turns out to be true. This scenario, for instance, leads Mavis to check the security guard's records and show that a panel truck bearing Maryland license plate number MDC-578 was in the STEMQ parking area on the day before the discovery that the cesium-137 canister was missing.

Fusing all the discovered evidence, Mavis concludes that it is very likely that the cesium-137 canister was stolen with the MDC-578 truck. After further investigation, she concludes that the two competing hypotheses, "misplaced" and "used in a project," lack evidentiary support.

8.3.5 **Hypothesis Specialization**

Continuing the investigation with the rental company owning the truck, she discovered that Omar al-Massari rented the MDC-578 truck, giving his alias, Omer Riley, a false address and that the truck is now contaminated because cesium-137 is radioactive. These lead to the conclusion that Omar used the truck to steal the cesium-137 canister. She further discovered that Omar al-Massari has ties with terrorist organizations and that he has given the cesium-137 canister to Saeed al-Nami, alias Kenny Derwish, who is a member of the terrorist organization Jihad Bis Sayf. These discoveries lead to the specializations of the hypotheses, as shown in Figure 6.

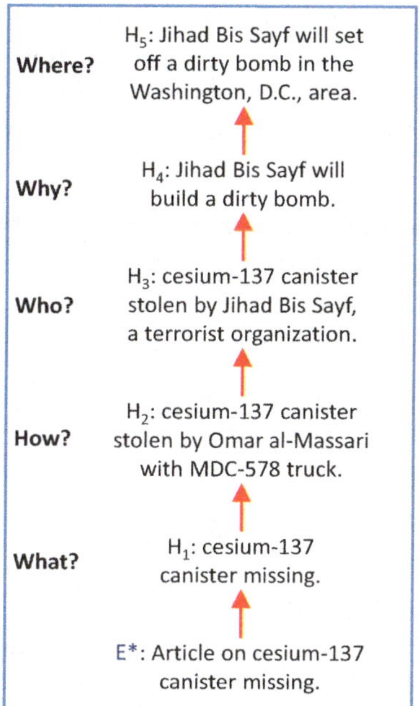

Figure 47. Hypothesis specialization.

Figure 49 shows the analysis of the top anticipatory hypothesis, "Jihad Bis Sayf will set off a dirty bomb in the Washington, DC, area."

It shows that Jihad Bis Sayf has the reasons, desire, and capability to set off the dirty bomb. It has reasons because it is a terrorist organization opposed to the United States, it has a presence in the Washington, DC, area, and a dirty bomb in this area would have a very high impact. Furthermore, Jihad Bis Sayf has the ability to build the bomb and set it off. In particular, it has the radioactive material from the stolen cesium-137 canister, and further investigation has determined that it has both the necessary explosive material (Saeed al-Nami has stolen 2 pounds of RDX explosive) and expertise to build the bomb (Saeed al-Nami has expertise in explosives and has received training in the building of dirty bombs).

Through such spiral hybrid reasoning, where abductions, deductions, and inductions feed on each other in recursive calls, Mavis continuously generates and updates intermediate alternative hypotheses, uses these hypotheses to guide the collection of relevant evidence, and uses the evidence to test these hypotheses, until the probability of the top-level hypotheses are assessed, ultimately anticipating that Jihad Bis Sayf will likely set off a dirty bomb in the Washington, D.C., area.

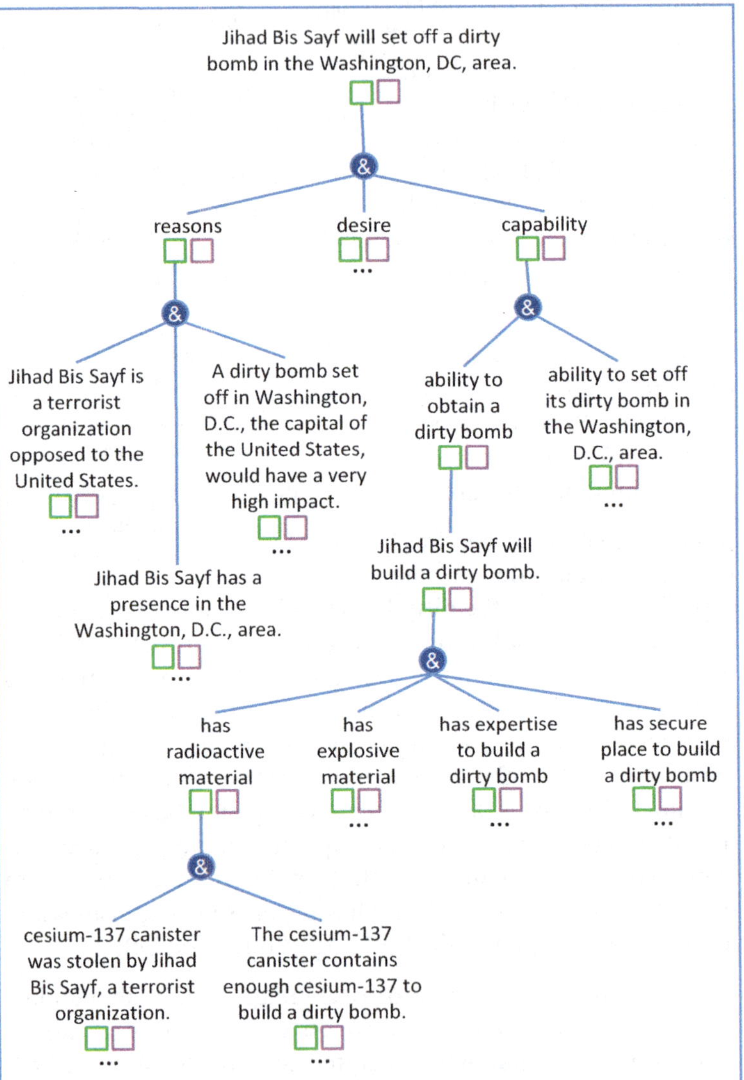

Figure 48. Anticipatory analysis.

Note, however, that performing this analysis is not as simple as one may infer from this presentation. It is the presented methodology that guides the analyst and simplifies it. Many things can and will indeed go wrong. You come up with some hypotheses based on evidence, but then you cannot find more evidence to support any of them. So, you need to come up with other hypotheses and always consider alternative ones. The deduction-based decomposition approach guides you on how to look for evidence, but your knowledge and imagination also play a crucial role. As illustrated here, Mavis imagined a scenario where the cesium-137 canister was stolen with a truck. But let us assume that she did not find supporting evidence for this scenario. Therefore, Mavis has to imagine other scenarios. Maybe the cesium canister was stolen by someone working at the STEMQ Company, or perhaps it was stolen by Ralph, the warehouse administrator. The important thing is that each such scenario opens up a new line of investigation.

8.3.6 Further Readings

Tecuci, G., Marcu, D. (2021). A Framework for Deep Anticipatory Intelligence Analysis, The 2021 AAAI Fall Symposium "Cognitive Systems for Anticipatory Thinking - 3rd Wave Autonomy," Arlington, VA, November 4-6.

8.4 **Law**

Figue 49 shows the essential structural elements of proof in law. Observe that there are three species of hypotheses or probanda [matters to be proved]:

The Ultimate Probandum refers to the essential or major fact in issue in the situation or case of concern to you.

Immediately below the ultimate probandum is some number of *Penultimate Probanda*. These penultimate probanda specify exactly what substantive law says you must establish in order to also establish the ultimate probandum.

A third species of probanda are the *Interim Probanda*. Under each penultimate probandum, there are actually two arguments, the argument of the prosecution (under the green square) that tries to prove that the penultimate probandum is true, and the argument of the defense (under the pink square) that tris to prove that the penultimate probandum is false. This pattern continues recursively wih each intermidiate probanda, the prosecution trying to prove it, while de defense to undermine it.

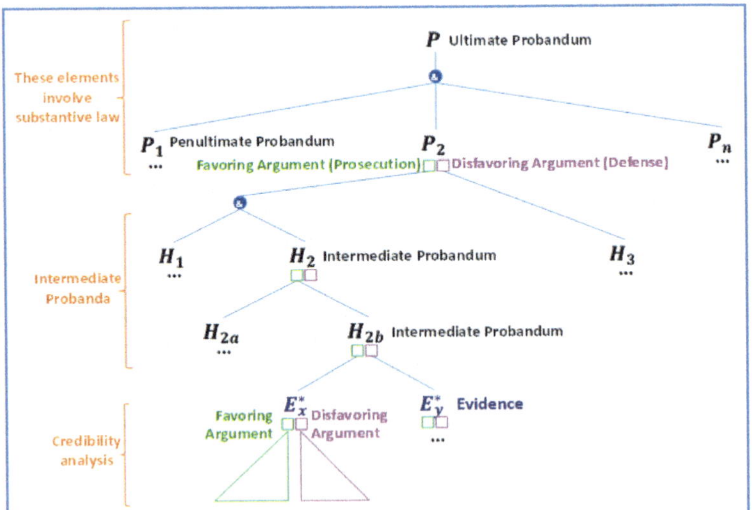

Figure 49. The essential structural elements of proof in law.

8.4.1 **Hypothesis Generation**

Let's consider a specific situation. You are a detective called to investigate the death of a victim whom we shall call Vic. The body of Vic was discovered on the floor of his garage. On the way to the scene, you entertain four initial hypotheses concerning the death of this person:

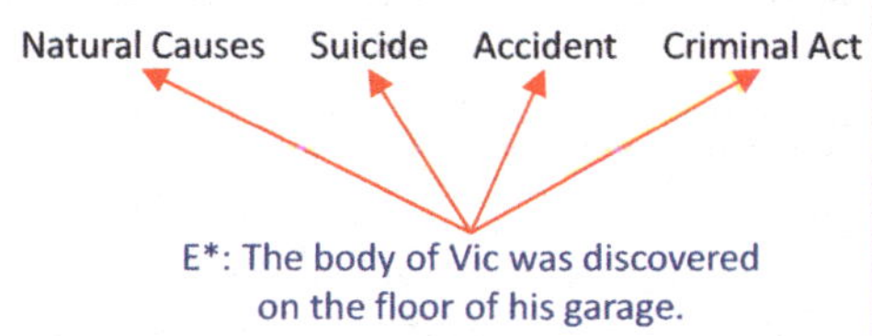

Figure 50. Initial hypotheses.

- Vic died of natural causes
- Vic committed suicide
- Vic's death was the result of an accident
- Vic died as the result of a criminal act

8.4.2 **Evidence Discovery**

Figure 51 is a general account of the sequence of events in the resolution of a legal dispute. The legal dispute begin during some Interval of Sustantive Importance (ISI) where an event or events happen that may subsequently give rise to some form of legal dispute. A process of fact investigation then takes place until the matter is brought to trial or some other form of settlement. At trial [or settlement] there is some episode of proof in which opposing parties present their evidence and arguments. If there is a trial, there is then some episode of deliberation and choice in which fact finders render a verdict. There may, of course, be other further activities such as appeals. Our present interests involve the process of fact investigation and the discovery-related processes it requires.

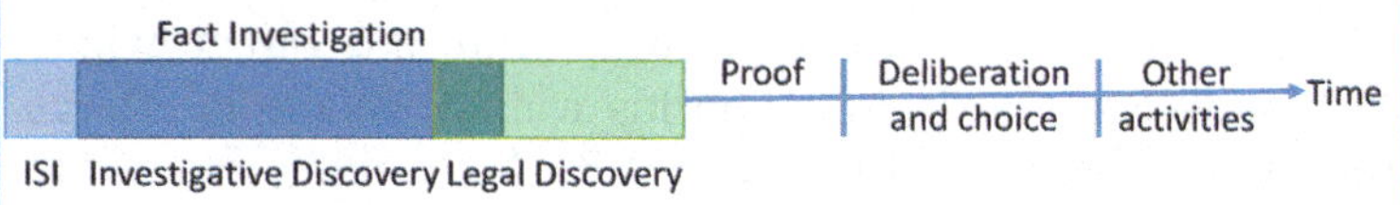

Figure 51. Sequence of events in the resolution of a legal dispute.

We have mentioned two possibly overlapping forms of discovery. Investigative discovery to indicate the proof-related ingredients a lawyer may generate on her/his own. By the term legal discovery we refer to the process of using legally-sanctioned rules for obtaining evidence and information from an opponent. Notice that we have shown investigative and legal discovery as overlapping.

We might regard your initial collection of hypotheses as your initial frame of discernment (Shafer, 1976). Observe that all of these initial hypotheses or possibilities have a common property: they are all vague, diffuse, or undifferentiated. For example, all the Criminal Act hypohesis says is that Vic died of a criminal act; it doesn't specify such matters as: what kind of criminal act it was, who did it, when it was done, or why it was done.

When you arrive at the scene, you begin to ask questions, make observations, and thus gather evidence. The first thing you observe is that Vic is lying in a pool of blood and has massive head injuries. Further evidence you collect at the scene and from forensic examiners allows you to rule out all possibilities but the Criminal Act one.

But now you gather the following evidence and have the first basis for refining this hypothesis or, if you like, making it more specific or less undifferentiated. There is no wallet found on Vic's body; his wife tells you that Vic usually carries it with him wherever he goes. In light of this evidence, you now entertain the possibilities:

- Vic was assaulted as a concomitant of robbery,
- The assault on Vic was made to appear as a concomitant of robbery,
- The assault on Vic was not connected with robbery (he was actually not carrying his wallet at all).

But now, under a bush not far away from the garage where Vic's body was found, you find his wallet containing a large sum of money and all of his credit cards. Nothing appears to have been taken from it. So, you begin to focus on H_{42}, that Vic was killed by a person who tried to make the assault look like it occurred during a robbery attempt.

You keep looking. The next item of evidence you find is a small packet of cocaine in the inside pocket of the jacket Vic was wearing. Was Vic's death the result of a drug deal gone sour? You now contemplate the following possibilities:

- Vic, a user of narcotics, was killed during an assault on him resulting from a drug deal gone sour, where the assault was made to look like it occurred during a robbery attempt.
- Vic, a pusher of narcotics, was killed during an assault on him resulting from a drug deal gone sour, where the assault was made to look like it occurred during a robbery attempt.

There are other possibilities at this level of discernment that we need not consider. Now, from a witness who appears at the scene, you learn that a white male was observed in the garage with Vic, just a short time before Vic was assaulted. So, you could refine to read:

- Vic, a pusher of narcotics, was killed during an assault on him by a white male that resulted from a drug deal gone sour, where the assault was made to look like it occurred during a robbery attempt.

In this example, we have the following important consequences.

First, evidence we collect and believe gets incorporated into our hypotheses; this is how an intially undifferentiated or diffuse hypothesis gets refined or becomes more specific.

Second, notice that each successive refinement is a conjunctive proposition containing one or more new events in the conjunction. We start out with A; then we have A & B; then we have A & B & C; and so on.

Third, a hypothesis at one level is simply a refinement of, and entirely consistent with, some hypothesis at the next higher [more diffuse] level. For example, says the same thing as H plus additional details.

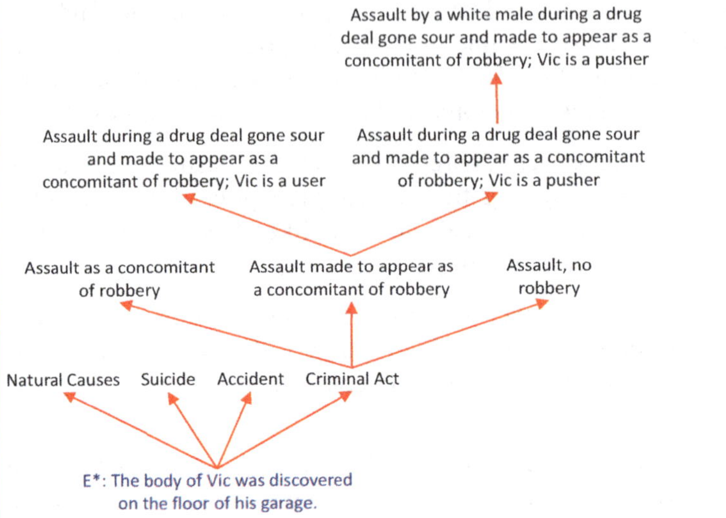

Figure 52. Sequence of events in the resolution of a legal dispute.

Fourth, and most important for our present purpose, the more refined a possibility or hypothesis becomes, the more it begins to resemble a story or a scenario.

But these refined hypotheses also have a most important characteristic: they never outrun the evidence we have. This is what distinguishes a refined hypothesis from a story or scenario. Here is the essential reason.

Stories, narrative, scenarios, or whatever you wish to call them, are always a mixture of "fact" and fancy. As William Twining notes (2001), stories have a beginning and an ending; they also have plots. There is often a temporal element to a story, scenario, or narrative. Events occur over time; one leads to another. So we often, in telling a story, attempt to show how one event may have given rise to or caused another. Many stories, narratives, or scenarios attempt causal explanations of things.

Legal dispute resolution is the only context known to us in which the processes of discovery, proof, and choice are more or less well-ordered in the manner shown in Figure 53. In all other contexts known to us, the process of discovery is ongoing; all the while there are cycles of other activities such as proof and choice, as indicated in Figure 54.

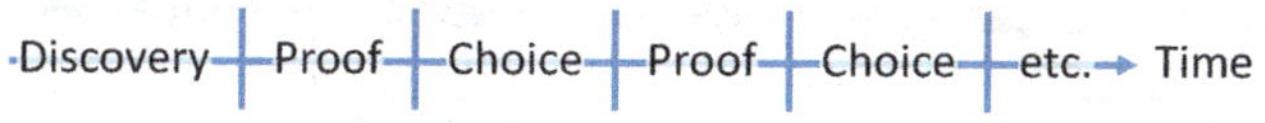

Figure 53. The process of discovery.

8.4.3 Hypothesis Testing

Law is the only domain known to me where the top level of the argumentation is provided. For example, to prove first degree murder of Vic, one needs to prove the four probanda from Figure 54.

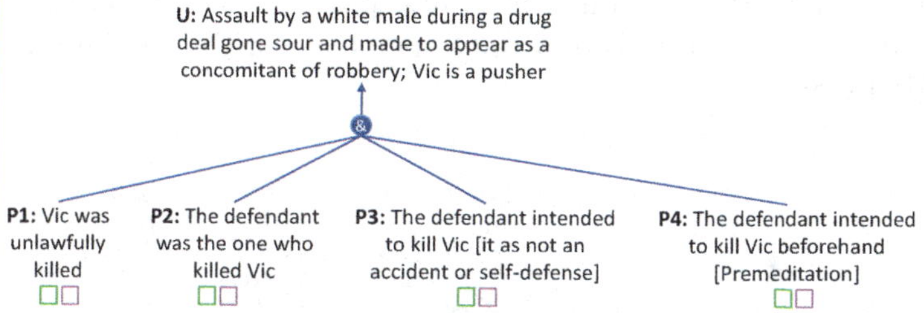

Figure 54. Example of a Wigmorean network.

8.5 Forensics

Observations made at the site of an explosion in a power plant lead to the formulation of several possible causes. Analysis of each possible cause leads to the discovery of new evidence that eliminates or refines some of the causes, and may even suggest new ones. This cycle continues until enough evidence is found to determine the most likely cause (see Figure 55).

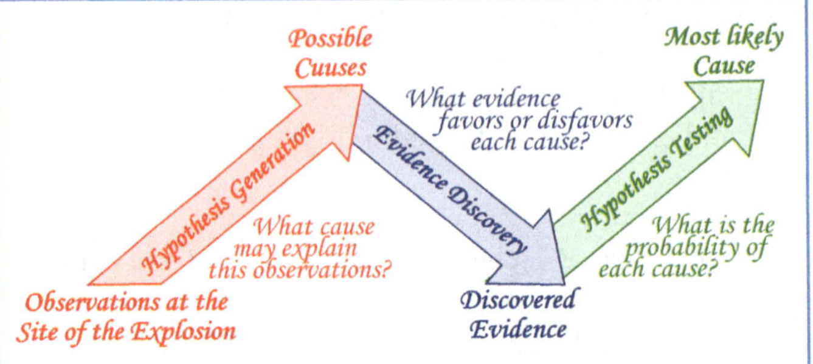

Figure 55. Critical thinking in forensics.

8.6 Medical Diagnosis

A doctor generates possible diagnoses that would explain a patient's complaints or doctor's observations. The doctor then performs various medical tests that provide further evidence that can be used to form a final diagnosis for the patient (see Figure 56).

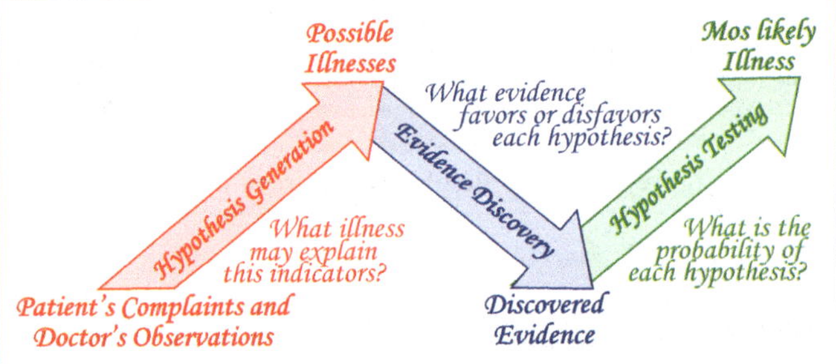

Figure 56. Critical thinking in medicine.

9 EPILOGUE

9.1 The Weaknesses of Democracy

Socrates
c.470–399BC

Socrates criticized Athenian democracy for allowing selfish individuals to gain power and wealth by using speech-making tricks and flattery to gain the support of citizens.

Plato
c.427–348BC

Plato also emphasized the risks of bringing dictators, tyrants, and demagogues to power, that democracies have leaders without proper skills or morals, and that it is quite unlikely that the best equipped to rule will come to power.

Mhai Eminescu
1850-1889

Eminescu warned that politicians' mistakes are crimes because millions of innocent people suffer. As a result, the development of an entire country is hindered, and its future is hindered for decades to come.

9.2 The American Democracy

Alexander Hamilton
1757–1804

The founding fathers of American Democracy understood well both the limits of democracy and the human nature. In the Federalist Papers, ***Alexander Hamilton*** wrote: "*If men were angels, no government would be necessary. If angels were to govern men, neither external nor internal controls on government would be necessary.*"

George Mason
1725-1792

George Mason started the *Virginia Declaration of Rights* with: "*That all men are by nature equally free and independent and have certain inherent rights, of which, when they enter into a state of society, they cannot, by any compact, deprive or divest their posterity; namely, the enjoyment of life* *and liberty, with the means of acquiring and possessing property and pursuing and obtaining happiness and safety."*

Thomas
Jefferson
1743-1826

Thomas Jefferson rephrased it memorably in the opening paragraphs of the *Declaration of Independence*: *"We hold these truths to be self-evident, that all men are created equal, that they are endowed by their Creator with certain unalienable Rights, that among these are Life, Liberty and the pursuit of Happiness."*

Martin Luther
King,Jr
1929-1968

Martin Luther King Jr. added: *"I have a dream that my four children will one day live in a nation where they will not be judged by the color of their skin but by the content of their character."*

Universal, in its appeal, the Virginia Declaration of Rights is the basis of the *U.S. Bill of Rights*, the *French Declaration of the Rights of Man and the Citizen* of 1789, and the *Declaration of Human Rights of the United Nations*.

The system of government put in place by these visionaries, based on checks and balances, equal status of the citizens in front of the law, and support of meritocracy, became an example for all nations. It attracted many of the best and brightest from around the world to pursue the American dream and transformed the United States into the most powerful and prosperous nation that ever existed.

9.3 **Critical Thinking and Wisdom**

Democracy was always vulnerable because many people are not rational decision makers. They elect leader based on impressions and emotions, rather than their real qualities. As a result, celibrities like actors or champions are vrey influential.

Today, our democracy is in more danger than ever because of a dishonest press that cover coruption and enable politicans to lie with impunity.

This book was written to help you become a better critical thinker, judge sources to dermine wewheer they could be trusted, judge statements, recognize fake news, and form your own views rather than adopting the views of others.

Critical thinking is the foundation of wisdom, and we end this book with the words of some of wisest people that ever existed.

Arsenie Papacioc

Arsenie Papacioc
1914-2011

The Orthodox Church unceasingly testifies that without Christ we can do nothing, for from Him and through Him we have life. Therefore, without an authentic life in Christ, man still dies alive. This was said by the Saviour Jesus Christ Himself: "Abide in Me and I in you. As the weeds cannot bear fruit of themselves unless they remain in the vine, so neither can you unless you remain in Me. I am the vine, you are the wheat. He who abides in Me and I in him brings forth much fruit, for without Me you can do nothing. If anyone does not abide in Me, he casts himself out as a swamp and withers; and he gathers them up and throws them into the fire and they burn. If you abide in me and my words abide in you, ask for what you want and it will be given to you" (John 15:4-7).

Cicero

Cicero
106-43BC

- *Gratitude is not only the greatest of virtues, but the parent of all others.*
- *Where is there dignity unless there is honesty?*

Marcus Aurelius, Meditations

- *When you arise in the morning think of what a privilege it is to be alive, to think, to enjoy, to love ...*
- *The chief task in life is simply this: to identify and separate matters so that I can say clearl....y to myself which are externals not under my control, and which have to do with*

the choices I actually control. Where then do I look for good and evil? Not to uncontrollable externals, but within myself to the choices that are my own...Very little is needed to make a happy life; it is all within yourself in your way of thinking.

Marcus Aurelius 121-180

- *Be content with what you have; rejoice in the way things are. When you realize there is nothing lacking, the whole world belongs to you.*
- *If it is not right do not do it; if it is not true do not say it.*
- *The best revenge is to be unlike him who performed the injury.*
- *If someone is able to show me that what I think or do is not right, I will happily change, for I seek the truth, by which no one was ever truly harmed. It is the person who continues in his self-deception and ignorance who is harmed.*

Seneca 4 BC – AD 65

Seneca

- *It is not the man who has too little, but the man who craves more, that is poor.*
- *It if not that we have little time,but more that we waiste a good deal of it.*

Gaius Musonius Rufus

If you accomplish something good wih hard work, the labor passes quickly, but the good endures; if you do something shamefull in pusuit of pleasure, the passes quickly, but the shame endures.

Gaius Musonius Rufus c. AD20-30, 101

Sun Tzu

- *Simplicity, patience, compassion. These three are your greatest treasures.*
- *Be content with what you have; rejoice in the way things are. When you realize there is nothing lacking, the whole*

world belongs to you.

- *A journey of a thousand miles must begin with a single step.*

Sun Tzu
544-496BC

Japenese Proverb

- The bambo that bends is strotogr that the oak that ressits.

Thomas Edison

Thomas Edison
1847-1931

- *Many of life's failures are people who did not realize how close they were to success when they gave up.*

God Bless!

GLOSSARY OF TERMS

Accuracy. Also termed *sensitivity.* An attribute of the credibility of certain kinds of tangible evidence such as those provided by sensing devices and tabled information.

Ambiguous evidence. Evidence which is imprecisely stated and we cannot determine exactly what it is telling.

Analysis. A reasoning operation by which we break down a hypothesis into components to better assess it. This operation is complementary to synthesis.

Argument. A chain of reasoning connecting observed evidence with a hypothesis of interest; The links in such chains represent ordered sequences of sources of doubt the analyst believes to be interposed between evidence and hypothesis.

Assumption. Statement taken to be likely true, but without supporting evidence, are called assumptions.

Authenticity. An attribute of the credibility of tangible evidence referring to whether a tangible item is what it is claimed to be.

Baconian probability system. A probability system based on Sir Francis Bacon's eliminative and variative views on evidential reasoning. This is the only system that captures how complete our evidence covers matters that should be covered in an inference problem.

Balance of probability. A probability standard used in law to refer to cases in which the evidence just favors one hypothesis over another even by the smallest amount.

Bayesian probability system. A probability system based on the conventional axioms of probability that concern games of chance and statistics, both of which involve repeatable phenomena. This system has many deficits when applied to cases involving non-repeatable events and situations.

Belief functions. A probability system very useful for non-repeatable events and situations. In this system beliefs are associated with any combination of the considered hypotheses.

Credibility. Credential of evidence indicating the degree to which we can believe what the evidence is telling us.

Beyond reasonable doubt. A probability standard used in law to refer to the highest grade of support evidence can provide some hypothesis of interest.

Competence. Credibility credential of human sources of evidence. A source of evidence is competent if she/he had access to what was reported and was able to understand it. Competence also refers to any skills a person might have to do some required job.

Conclusive evidence. If believable, such evidence would make some conclusion certain.

Conjunction. The combination of judgments about the probability of individual hypotheses into a single hypothesis as a whole, where all the individual hypotheses need to be true to make the single hypothesis true. See also Disjunction.

Connecting the dots. The task of marshaling thoughts and evidence in the generation or discovery of productive hypotheses and new evidence, and in the construction of defensible and persuasive arguments on hypotheses we believe to be most favored by the evidence we have gathered and evaluated.

Contradictory evidence. Type of dissonant evidence involving events that are mutually exclusive, i.e. they cannot occur jointly. For example, one evidence item says that event E occurred and another evidence item says that event E did not occur.

Convergent evidence. Two or more evidence items that concern different events which point toward or favor the same hypothesis

Corroborative evidence. Evidence that reports the same event.

Credential of evidence. A term used to describe a property of evidence that needs to be established or justified. Three major credentials of evidence are: relevance, credibility or credibility, and inferential force or weight.

Credibility. Concerns the extent to which an item of evidence or a source of evidence may be believed. On occasion, this term is wrongly equated with the term reliability (q.v., which has a more restricted definition). As a credential of evidence, credibility has several different attributes that depend upon the form of evidence, whether it is tangible or testimonial.

Credibility attributes. For tangible evidence, these attributes are: authenticity, accuracy, and reliability. For testimonial evidence, these attributes are: veracity, objectivity, and observational sensitivity.

Critical reasoning. Reasoning represented by an argument that is logically coherent, free of disconnects and non sequiturs.

Data. Refer to un-interpreted signals, raw observations, or measurements, such as such as the number 6 or the color red. See also Evidence and data or item of information.

Deductive reasoning. A form of reasoning that makes some conclusion necessarily true or certain.

Defensible and persuasive argument. An argument is defensible if it is free from logical disconnects. It is persuasive if it is compelling. The trouble is that not all persuasive arguments are defensible and not all defensible arguments persuasive.

Demonstrative tangible evidence. Evidence not of a thing itself but of a representation or image of this thing.

Discovery. Refers to the process of generating new hypotheses or new lines of inquiry and new evidence.

Disjunction. The combination of judgments about the probability of individual hypotheses into a single hypothesis as a whole, where only one of the individual hypotheses need to be true to make the single hypothesis true.

Disfavoring evidence. Evidence that argues against the truth of some hypothesis.

Dissonant evidence. Directionally inconsistent items of evidence pointing toward different hypotheses.

Divergent evidence. Type of dissonant evidence where the evidence points to different hypotheses as opposed to contradictory evidence which involves events that are mutually exclusive.

Divide and conquer. The act of decomposing a complex reasoning task into its simpler ingredients and of combining their conclusions. See Analysis and Synthesis.

Dots. Details in the observable information or data about an intelligence situation, as well as potential links in chains of reasoning or arguments we may construct to link dots to hypotheses we are trying to prove or disprove.

Epistemology. A branch of philosophy concerning the acquisition of and validity of knowledge.

Evidence. Evidence is any observable sign, indicator, or datum we believe is relevant in deciding upon the extent to which we infer any hypotheses we have entertained as being correct or incorrect.

Evidence and data or item of information. Evidence differs from data or items of information. Data or Items of information only become evidence when their relevance is established regarding some matter to be proved or disproved.

Evidence and events. There is an important distinction to be made between evidence of some event and the event itself. Having evidence that an event occurred does not entail that this event did occur. What is at issue is the credibility of the evidence and its source(s).

Evidence-based hypothesis assessment. The process of determining the probability of a hypothesis based on the available evidence.

Evidential dots. One of two forms of dots that must be connected. The other form of dots concern ideas about the meaning of an evidential dot.

Evidentiary testing of hypothesis. See Evidence-based hypothesis assessment.

Fact. Any event or act or condition of things, assumed as having happened or having existed.

Favoring evidence. Evidence that is directionally consistent in favoring the same hypothesis.

Force. See Inferential force or weight.

Fuzzy probability system. A probability system where the uncertainty about a conclusion reached is expressed in words (such as "likely" or "almost certain"), each "fuzzy" word being related to a range of numerical probabilities by a possibility function.

Harmonious evidence. Two or more items of evidence that are directionally consistent in the sense that they all point toward, or favor, the same hypothesis or possible conclusion.

Heuristic. A rule of thumb that aids you in any discovery, inference, learning, or decision problem.

Holistic approach to analysis. Work on an analysis problem where you do all in your own head without decomposing it in any way or seeking the assistance of others.

HUMINT. Testimonial evidence given by a human source about some matter of intelligence interest.

Hypothesis. A general proposition put forward as a possible explanation for known facts from which additional investigations can be planned to generate evidential data that will tend to strengthen or weaken the basis for accepting the proposition as the best or strongest explanation of the available data. Hypotheses commonly refer to possible alternative conclusions we could entertain about matters of interest in an analysis.

Inconclusive evidence. Evidence which is consistent with the truth of more than one hypothesis or possible explanation.

Inference. The process of deriving logical conclusions from premises.

Inference network. Network consisting of multiple lines of argument that connect many different kinds of evidence to the hypothesis under consideration.

Inferential force or weight. Credential of evidence indicating how strong the evidence is in favoring or disfavoring hypotheses we are considering.

Information. Data equipped with meaning provided by a certain context, such as "6 a.m." or "$6". See also Evidence and data or item of information.

Knowledge. Justified true belief. We say that Person *A* knows that event B occurred if the event B did occur (true), the person *A* got non-defective evidence that B occurred (justified), and *A* believed this evidence (belief).

Marshaling. Refers to the bringing together of thoughts and evidence during hypotheses generation and argument construction. Having useful strategies for marshaling thus helps advance the processes of hypotheses generation and analysis.

Nugget. A term used by intelligence agencies with reference to believable/credible evidence that would make some conclusion certain.

Objectivity. An attribute of the credibility of a human source characterizing the extent to which that source based her/his belief that the reported belief occurred on her/his sensory evidence rather than on what this source expected or desired to observe.

Observational sensitivity. Attribute of the credibility of a human source characterizing how good was the sensory evidence this source received under the conditions in which her/his observation was made.

Probability. Characterizes the uncertainty about a given event, statement, hypothesis or conclusion. Differing conceptions of probability are a matter of considerable controversy and debate within statistics and the logic of proof.

Proposition. A statement that is true or false, that can be affirmed or denied.

Relevance. Credential of evidence indicating how a datum or information item is linked to something we are trying to prove or disprove.

Relevant evidence. Evidence having any tendency to make the existence of any fact that is of consequence to the determination of the action more probable or less probable than it would have been without the evidence. Federal Rule of Evidence 401.

Reliability. Reliability is especially relevant to various forms of sensors that provide us with many forms of demonstrative tangible evidence. A system, sensor, or test of any kind is reliable to the extent that the results it provides are repeatable or consistent. This term is often used incorrectly as a synonym for the term *credibility* or *credibility,* which involves other attributes.

Tangible evidence. Evidence that can be directly examined by persons drawing conclusions to see what event(s) this evidence reveals. Examples include objects, documents, images, measurements, and charts.

Testimonial evidence. Evidence provided by a human source. Testimonial evidence about some event can be based on direct observations, secondhand reports from another source, or on the basis of opinion or inferences based on information about the occurrence of other events.

Trifles. A term used by Sherlock Holmes to refer to the many details (dots) he observed that formed the basis for his investigations.

Understandability. An attribute of the competence of a human source characterizing the extent to which that source understood what was being observed well enough to provide us with an intelligible account.

Veracity. An attribute of the credibility of a human source characterizing the extent to which that source believes that the reported event occurred.

REFERENCES

Anderson, T., Schum, D., and Twining, W. (2005). Analysis of Evidence, Cambridge University Press.

Cohen, L. J. (1977). The probable and the provable. Oxford, UK: Clarendon Press.

Danzig, R., Sageman, M., Leighton, T., Hough, L., Yuki, H., Kotani, R., and Hosford Z.,M. (2011). Aum Shinrikyo: Insights Into How Terrorists Develop Biological and Chemical Weapons, Center for a New American Security, Washington, DC, July.

Drogin, B. (2007). CURVEBALL: Spies, Lies, and the Con Man Who Caused a War. Random House, New York, NY.

ICD 203. Intelligence Community Directive 203, 21 June 2007.

Kolmogorov, A. N. (1956). Foundations of a Theory of Probability (1933), 2nd English edition, Chelsea Publishing, New York, NY., pp. 3-4.

Marrin, S. (2011). Improving Intelligence Analysis: Bridging the gap between scholarship and practice. Routledge, London, and New York.

MacEachin D., Forward to Heuer R.J., Psychology of Intelligence Analysis, Center for the Study of Intelligence, Central Intelligence Agency, 1999. http://www.au.af.mil/au/awc/awcgate/psych-intel/art2.html

Peirce, C. S. (1901). Abduction and Induction, in Buchler, J. (ed.) (1995). Philosophical Writings of Peirce, 150-156, Dover: New York, NewYork.

Schum, D. A. (2001). The evidential foundations of probabilistic reasoning. Evanston, IL: Northwestern University Press.

Scriven, M., and Paul, R. (1987). A statement presented at the 8th Annual International Conference on Critical Thinking and Education Reform, Summer 1987. Retrieved February 21, 2022 from http://www.criticalthinking.org/pages/defining-critical-thinking/766.

Shafer, G. (1976). A Mathematical Theory of Evidence, Princeton University Press, Princeton, NJ.

Tecuci, G., Boicu, M., Marcu, D., Boicu, C., Barbulescu, M. (2008a). Disciple-LTA: Learning, Tutoring and Analytic Assistance, Journal of Intelligence Community Research and Development (JICRD), July. http://lac.gmu.edu/publications/2008/Disciple-LTA08.pdf

Tecuci, G., Schum, D. A., Boicu, M., Marcu, D. (2011). Introduction to Intelligence Analysis: A Hands-on Approach with TIACRITIS, 220 pages, Learning Agents Center, George Mason University, First edition 2010, Second edition 2011. http://lac.gmu.edu/publications/2011/TecuciG_Introduction_Intelligence_Analysis.pdf

Tecuci, G., Schum, D.A., Marcu, D., Boicu, M. (2016a). Intelligence Analysis as Discovery of Evidence, Hypotheses, and Arguments: Connecting the Dots, Cambridge University Press. http://lac.gmu.edu/IABook/

Tecuci, G., Marcu, D., Boicu, M., Schum, D. A. (2016b). Knowledge Engineering: Building Cognitive Assistants for Evidence-based Reasoning, Cambridge University Press. http://lac.gmu.edu/KEBook/

Twining W. (2003). Evidence as a Multi-Disciplinary Subject, Law, Probability and Risk, 2, 91-107, 2003.

Wigmore, J. H. (1937). The Science of Judicial Proof: As Given by Logic, Psychology, and General Experience, and Illustrated in Judicial Trials. 3rd Edition, Little, Brown & Co., Boston, MA, 1937.

Illustration Credits

(by the page the sources of the illustration used in this book)

Cover Adapted from
https://en.wikipedia.org/wiki/Thinker_of_Hamangia#/media/File:Bucharest_-_The_Thinker_&_The_Sitting_Woman_of_Cernavoda_-_no_bg.png
which is used under CC BY-SA 4.0 License:
https://creativecommons.org/licenses/by-sa/4.0/

16 https://en.wikipedia.org/wiki/Aristotle
16 https://en.wikipedia.org/wiki/Galileo_Galilei
16 https://en.wikipedia.org/wiki/Isaac_Newton
16 https://en.wikipedia.org/wiki/William_Whewell
16 https://en.wikipedia.org/wiki/John_Locke
16 https://en.wikipedia.org/wiki/John_Henry_Wigmore
16 https://en.wikipedia.org/wiki/Charles_Sanders_Peirce
24 https://en.wikipedia.org/wiki/Confucius
31 https://en.wikipedia.org/wiki/Marcus_Aurelius
52 https://lahora.gt/lh-suplementos-culturales/la-hora/2022/08/19/albert-einstein/
58 https://en.wikipedia.org/wiki/Pierre-Simon_Laplace
73 https://en.wikipedia.org/wiki/John_Henry_Wigmore
83 https://en.wikipedia.org/wiki/Francis_Bacon
102 https://en.wikipedia.org/wiki/Benjamin_Franklin
117 https://en.wikipedia.org/wiki/Jacques_Hadamard
147 https://en.wikipedia.org/wiki/Socrates
147 https://en.wikipedia.org/wiki/Plato
147 https://en.wikipedia.org/wiki/Mihai_Eminescu
148 https://en.wikipedia.org/wiki/George_Washington
148 https://en.wikipedia.org/wiki/George Mason
148 https://en.wikipedia.org/wiki/Thoms_Jefferson
148 https://en.wikipedia.org/wiki/Alexander_Hamilton
148 https://en.wikipedia.org/wiki/Martin_Luther_King_Jr.
149 https://en.m.wikipedia.org/wiki/File:Arsenie_Papacioc_by_Paul_Mecet.jpg
150 https://en.wikipedia.org/wiki/Cicero
150 https://en.wikipedia.org/wiki/Marcus_Aurelius
150 https://en.wikipedia.org/wiki/Seneca
150 https://en.wikipedia.org/wiki/Gaius_Musonius_Rufus
151 https://en.wikipedia.org/wiki/Sun_Tzu
151 https://en.wikipedia.org/wiki/Thomas_Edison

Gheorghe Tecuci is a Professor of Computer Science and Director of the Learning Agents Center in the School of Computing of George Mason University, a Member of the Romanian Academy, and former Chair of Artificial Intelligence in the Center for Strategic Leadership of the U.S. Army War College.

"Critical Thinking is a timely contribution to our field of AI, written by a true expert on this topic, that describes the foundations and methods for creating and assessing hypotheses based on interactively collected evidence. I recommend this for students (as well as curious researchers and practitioners) who wish to learn more about the underlying reasoning mechanisms and related topics it explores."

-David W. Aha, Director, Navy Center for Applied Research in AI, Naval Research Laboratory, Washington, DC, USA.

"This book is extremely important. In the present times, where humans are conducted by generative AI, it is crucial to reinforce human reasoning to build and accumulate correct and pertinent knowledge. This book represents a knowledge stone in human learning."

-Claude Frasson, Honorary and Associate Professor, University of Montreal, Quebec, Canada.

Made in the USA
Monee, IL
03 November 2026

67376104R00095